Digital Techniques

KV-322-952

071729

MATTHEW BOULTON
COLLEGE LIBRARY

Digital Techniques

MATTHEW
BOULTON
COLLEGE

K. F. Ibrahim

Senior Lecturer, Willesden College of Technology

Longman
Scientific &
Technical

Longman Scientific & Technical

MATTHEW
BOULTON
COLLEGE
LIBRARY

40061
621·3815

Longman Scientific & Technical,
Longman Group UK Limited
Longman House, Burnt Mill, Harlow,
Essex CM20 2JE, England
and Associated Companies throughout the world.

© Longman Group UK Limited 1991

All rights reserved; no part of this publication
may be reproduced, stored in a retrieval system,
or transmitted in any form or by any means, electronic,
mechanical, photocopying, recording, or otherwise
without either the prior written permission of the
Publishers or a licence permitting restricted copying in
the United Kingdom issued by the Copyright Licensing
Agency Ltd, 33–34 Alfred Place, London, WC1E 7DP.

First published 1991

British Library Cataloguing in Publication Data
Ibrahim, K. F.
 Digital techniques.
 1. Digital electronic equipment
 I. Title
 621.3815

Produced by Longman Singapore Publishers (Pte) Ltd
Printed in Singapore

ISBN 0-582-03688-7

Contents

Preface

This book covers the topic of digital electronics and techniques from basic principles (number systems) through simple logic elements (gates, registers and counters) to LSI and VLSI devices such as memory chips, microprocessors, peripheral devices and their interfaces. An introduction to microprocessor-based systems is provided in Chapter 9. The purpose of this is to provide the reader with a basic knowledge of microprocessor systems. Such systems incorporate a variety of digital devices and a grasp of the principles of these devices is essential to a proper understanding of digital electronics.

Logical fault finding procedures for gates and other digital circuits are covered in Chapter 12 together with a description of all relevant instruments such as the logic probe and pulser. Explanations of the operation and application of instruments such as the logic and signature analysers used for testing microprocessor-based systems are also provided.

The book covers the syllabus for City and Guilds 224 part 2 and 3 Digital Electronics as well as BTEC levels II and III. It is also suitable for other students and practising electronic engineers.

K. F. Ibrahim

MATTHEW BOULTON
COLLEGE LIBRARY

1 Number systems

We are used to a denary or decimal system of numbers employing ten different symbols, namely 0, 1, 2, ..., 9; i.e. a system with a base of 10. However this is not necessarily the best system for all applications. In digital electronics the simpler binary system is more suitable. This system has a base of 2, i.e. it uses two symbols only: 0 and 1. Other systems that may be used are the base-8, octal, and the base-16, hexadecimal, systems.

Denary and binary

In the denary system, the first column, A (Table 1.1) is the ones or units, the second column, B, is the tens, C is the hundreds, and so on. Columns A, B and C represent ascending powers of 10, namely $10^0 = 1$, $10^1 = 10$, $10^2 = 100$. Similarly for a binary system with a base of 2, each column represents a power of 2, namely $2^0 = 1$, $2^1 = 2$, $2^2 = 4$, and so on.

Table 1.1

Decimal columns			Binary columns		
C	B	A	C	B	A
10^2	10^1	10^0	2^2	2^1	2^0
= 100	= 10	= 1	= 4	= 2	= 1
(hundreds)	(tens)	(ones)	(fours)	(twos)	(ones)

Each binary digit is known as a bit with the rightmost digit known as the **least significant bit**, LSB and the leftmost digit known as the **most significant bit**, MSB. Table 1.2 gives a list of denary (base 10) numbers and their corresponding binary (base 2) numbers.

To distinguish between different number systems, the base is shown as a subscript. For example 9_{10} denotes 9 (denary), and 01101_2 denotes 01101(binary), and so on. The subscript is normally omitted where it is clear what the numbering system is.

Examples of converting binary to denary are shown in Table 1.3.

Table 1.2

Decimal	Binary		
	C (4)	B (2)	A (1)
0	0	0	0
1	0	0	1
2	0	1	0
3	0	1	1
4	1	0	0
5	1	0	1
6	1	1	0
7	1	1	1

Table 1.3

Binary		Binary columns					Denary
	32	16	8	4	2	1	
1110	—	—	1	1	1	0	= 8 + 4 + 2 = 14
1011	—	—	1	0	1	1	= 8 + 2 + 1 = 11
11001	—	1	1	0	0	1	= 16 + 8 + 1 = 25
10111	—	1	0	1	1	1	= 16 + 4 + 2 + 1 = 23
110010	1	1	0	0	1	0	= 32 + 16 + 2 = 50

Conversion of denary to binary

Conversion of denary to binary may be carried out by a combination of intuition and trial and error. The denary number is split into a sum of powers of 2. Table 1.4 shows some examples of such conversions. Consider the number 22_{10}. It is less than 2^5 ($= 32$) and hence a 0 is placed in that column. However it is greater than 2^4 ($= 16$), hence a 1 is placed in that column. This leaves $22 - 16 = 6$ which is greater than 2^2 ($= 4$). Hence a 1 is placed in that column leaving $6 - 4 = 2$. This produces a 1 in the 2^1 ($= 2$) column. Since there is no remainder, a 0 is placed in the 2^0 ($= 1$) column giving the binary number 010110.

Table 1.4

Denary number	Binary columns						Binary number
	2^5	2^4	2^3	2^2	2^1	2^0	
	32	16	8	4	2	1	
15	0	0	1	1	1	1	001111
22	0	1	0	1	1	0	010110
45	1	0	1	1	0	1	101101
52	1	1	0	1	0	0	110100

An alternative method consists of dividing the denary number successively by 2 and noting the remainder each time. The remainder being either 0 or 1, forms the binary number with the last remainder representing the most significant bit. For example to convert 52_{10} into binary the following steps are taken:

$$\frac{52}{2} = 26 \text{ remainder } 0 \text{ LSB}$$

$$\frac{26}{2} = 13 \text{ remainder } 0$$

$$\frac{13}{2} = 6 \text{ remainder } 1$$

$$\frac{6}{2} = 3 \text{ remainder } 0$$

$$\frac{3}{2} = 1 \text{ remainder } 1$$

$$\frac{1}{2} = 0 \text{ remainder } 1 \text{ MSB}$$

The answer, therefore, is 110100_2.

The above method can be used for conversion to other numbering systems such as octal or hexadecimal.

EXERCISE 1.1

1. Convert the following binary numbers into denary.
   (a) 110 (b) 1110 (c) 10101 (d) 101101 (e) 111111

2. Convert the following denary numbers to binary.
   (a) 5 (b) 17 (c) 42 (d) 31 (e) 47

Octal numbers

Octal numbers have a base of 8 and as such have eight different symbols: 0, 1, 2, 3, 4, 5, 6, and 7. Octal columns in a number represent a power of 8, i.e.

column D	column C	column B	column A
8^3	8^2	8^1	8^0
= 521	= 64	= 8	= 1
(512s)	(64s)	(eights)	(units)

The successive division method of conversion may be employed to convert denary numbers to octal. The denary number is successively divided by the base 8 noting the remainder at each stage. For example to convert 5819_{10} to octal:

$$\frac{5819}{8} = 727 \text{ remainder } 3 \text{ LSB}$$

$$\frac{727}{8} = 90 \text{ remainder } 7$$

$$\frac{90}{8} = 11 \text{ remainder } 2$$

$$\frac{11}{8} = 1 \text{ remainder } 3$$

$$\frac{1}{8} = 0 \text{ remainder } 1 \text{ MSB}$$

Therefore $5819_{10} = 13273_8$.

Octal and binary

Each octal digit can be represented by a three-bit binary number (see Table 1.5). To convert octal to binary, each octal digit is converted separately. For instance 3527_8 may be converted as follows:

$$3_8 = 011_2 \text{ MSB}$$
$$5_8 = 101_2$$
$$2_8 = 010_2$$
$$7_8 = 111_2 \text{ LSB}$$

Hence $3527_8 = 011\ 101\ 010\ 111$ in binary.

Conversely, for binary to octal conversion, the binary number may be arranged into groups of three binary bits. Each group is then

Table 1.5

Octal	Binary
0	000
1	001
2	010
3	011
4	100
5	101
6	110
7	111

separately converted into octal. For instance 11110011001_2 is first arranged into groups of three bits starting from the least significant bit (right-hand side) to give 11 110 011 001 which is then converted:

$$11_2 = 3_8 \text{ most significant digit}$$
$$110_2 = 6_8$$
$$011_2 = 3_8$$
$$001_2 = 1_8$$

giving an octal number of 3631.

Table 1.6

Hexadecimal	Denary
0	0
1	1
2	2
3	3
4	4
5	5
6	6
7	7
8	8
9	9
A	10
B	11
C	12
D	13
E	14
F	15

EXERCISE 1.2

1. Convert the following octal numbers to denary.
   (a) 32_8 (b) 57_8 (c) 213_8 (d) 156_8

2. Convert the following denary numbers to octal.
   (a) 28_{10} (b) 137_{10} (c) 351_{10} (d) 629_{10}

3. Convert the following octal numbers to binary.
   (a) 27_8 (b) 210_8 (c) 555_8 (d) 6543_8

4. Convert the following binary numbers into octal.
   (a) 010 (b) 110011 (c) 1011001 (d) 1010111000

Hexadecimal numbers

Hexadecimal (normally referred to as hex) numbers have a base of 16 and as such have sixteen distinct symbols (Table 1.6).

Numbers above 15_{10} require more than one hex digit. Hexadecimal columns represent the power of 16, e.g. $16^0 = 1$ (units column), $16^1 = 16$, $16^2 = 256$, and so on. For instance:

$$152B_{16} = (1 \times 16^3) + (5 \times 16^2) + (2 \times 16^1) + (11 \times 16^0)$$
$$= 1 \times 4096 + 5 \times 256 + 2 \times 16 + 11 \times 1$$
$$= 4096 + 1280 + 32 + 11$$
$$= 5419_{10}.$$

Conversely to convert denary numbers to hexadecimal, the successive division method should be used as explained earlier. This time, however, the division is by the base number 16. For example, to convert 3409_{10} to a hexadecimal number:

$$\frac{3409}{16} = 213 \text{ remainder } 1_{10} = 1_{16} \text{ least significant digit}$$

$$\frac{213}{16} = 13 \text{ remainder } 5_{10} = 5_{10}$$

$$\frac{13}{16} = 0 \text{ remainder } 13_{10} = D_{16} \text{ most significant digit}$$

Hence $3409_{10} = D51_{16}$.

Hexadecimal and binary

Each hex number can be represented by a four-bit binary number as in Table 1.7.

Table 1.7

Hexadecimal	Binary
0	0000
1	0001
2	0010
3	0011
4	0100
5	0101
6	0110
7	0111
8	1000
9	1001
A	1010
B	1011
C	1100
D	1101
E	1110
F	1111

To convert hex numbers into binary, each digit is converted separately into the appropriate four-bit binary. For example $2A5C_{16}$ may be converted into binary as follows.

$2_{16} = 0010$ most significant
$A_{16} = 1010$
$5_{16} = 0101$
$C_{16} = 1100$ least significant

This gives a binary number of 0010 1010 0101 1100.

Conversely, binary numbers may be arranged into groups of four bits for easy conversion into hexadecimal. For instance 0100111101011110_2 may be grouped as 0100 1111 0101 1110. Converting each group separately we get:

$0100_2 = 4_{16}$ most significant
$1111_2 = F_{16}$
$0101_2 = 5_{16}$
$1110_2 = E_{16}$ least significant

giving a hex number of 4F5E.

EXERCISE 1.3

1. Convert the following hexadecimal numbers into binary.
   (a) 2A (b) 8D (c) C09 (d) EF2 (e) FFFF

2. Convert the following binary numbers into hexadecimal.
   (a) 11010110 (b) 110010 (c) 100101111111
   (d) 1110101100110101

3. Convert $3F1_{16}$ into a denary number.

Binary fractions

In the denary system, a fraction may be represented by the use of a decimal point. Denary digits to the left of the decimal point represent an ascending order and those to its right represent a descending order of the power of 10. Thus

$$0.1_{10} = 10^{-1} = \frac{1}{10}$$

$$0.01 = 10^{-2} = \frac{1}{10^2}$$

$$0.2 = 2 \times 0.1 = 2 \times 10^{-1}, \text{ and so on,}$$

and similarly, for the binary representation of fractions. A binary point is used to represent a descending order of negative powers of the base 2. Hence

$$0.1_2 = 2^{-1} = \frac{1}{2} \text{ and}$$

$$0.01_2 = 2^{-2} = \frac{1}{2^2} = \frac{1}{4}.$$

Table 1.8 shows a comparison between the denary and binary digit representation.

Table 1.8

			decimal pt			
10^2	10^1	10^0	↓	10^{-1}	10^{-2}	10^{-3}
hundreds	tens	ones	•	tenths	hundredths	thousandths
			binary pt			
2^2	2^1	2^0	↓	2^{-1}	2^{-2}	2^{-3}
fours	twos	ones	•	halves	quarters	eighths

It follows that

$$\text{binary } 0.111 = \tfrac{1}{2} + \tfrac{1}{4} + \tfrac{1}{8}$$
$$= 0.5 + 0.25 + 0.125$$
$$= 0.875 \text{ (denary)}.$$

Further

$$101.101_2 = 4 + 0 + 1 + \tfrac{1}{2} + 0 + \tfrac{1}{8}$$
$$= 5 + 0.5 + 0.125$$
$$= 5.625_{10}.$$

The conversion of decimal fractions into binary may be carried out by the successive multiplication of the decimal fraction by the base, 2, and using the whole number thus produced as a binary bit. The process is then repeated for the decimal fraction of the product, and so on. The first binary bit that is produced by this method is the most significant bit of the binary fraction. For example, to convert 0.625_{10} to a binary fraction:

$0.625 \times 2 = 1.25$, whole part = 1 (MSB), remainder = 0.25
$0.25 \times 2 = 0.5$, whole part = 0, remainder 0.5
$0.5 \times 2 = 1.0$, whole part = 1 (LSB) with no remainder.

Hence $0.625_{10} = 0.101_2$

EXAMPLE

Convert 25.125_{10} to binary.

Solution

Consider first the whole part of the denary number, namely 25.

$\frac{25}{2} = 12$; remainder 1 LSB

$\frac{12}{2} = 6$; remainder 0

$\frac{6}{2} = 3$; remainder 0

$\frac{3}{2} = 1$; remainder 1

$\frac{1}{2} = 0$; remainder 1 MSB,

giving $25_{10} = 11001_2$.

Now consider the fraction part of the denary number namely 0.125:

$$0.125 \times 2 = 0.25; \text{ whole part } = 0 \text{ MSB; remainder } = 0.25$$
$$0.25 \times 2 = 0.5; \text{ whole part } = 0; \text{ remainder } = 0.5$$
$$0.5 \times 2 = 1.0; \text{ whole part } = 1 \text{ LSB; with no remainder.}$$

This gives $0.125_{10} = 0.001_2$.
Hence $25.125_{10} = 11001.001_2$.

EXERCISE 1.4

1. Convert the following fractions to binary.
   (a) 0.25 (b) 0.21875 (c) 0.46875

2. Convert the following binary numbers to denary.
   (a) 0.01_2 (b) 0.11101_2 (c) 111.0111_2
   (d) 101001.01_2

Binary coded decimal (BCD)

So far we have considered the conversion of denary numbers into pure or natural binary. However it is very useful in certain applications, e.g. in microprocessor-based systems, to convert each denary digit separately into a four-bit binary code. A two-digit denary number will thus produce two groups of four-bit binary numbers (eight bits in total) regardless of the denary number itself. The result is known as **binary-coded decimal** (BCD). The code could be a direct conversion to binary known as the 8421 BCD, or some other encoding method. Table 1.9 shows two types of BCD systems.

The 8421 BCD is the most popular system of BCD encoding. Here each BCD digit is given the same weighting as for pure binary numbers, namely 8, 4, 2 and 1, starting from the most significant bit. For example

$$59_{10} = 0101\ 1001\ (8421\ BCD),$$
$$843_{10} = 1000\ 0100\ 0011\ (8421\ BCD).$$

In the 2421 BCD system, the MSB has a weighting of 2, the next

Table 1.9

Decimal number	8421 BCD	2421 BCD
0	0000	0000
1	0001	0001
2	0010	0010
3	0011	0011
4	0100	0100
5	0101	0101
6	0110	0110
7	0111	0111
8	1000	1110
9	1001	1111

MATTHEW BOULTON
COLLEGE LIBRARY

digit has a weighting of 4, followed by a weighting of 2 and finally the LSB has a weighting of 1. For example

$$59_{10} = 0101\ 1111\ (2421\ BCD),$$
$$843_{10} = 1110\ 0100\ 0011\ (2421\ BCD).$$

Binary arithmetic

Binary addition

Binary addition is similar to the addition of denary numbers. The two numbers are arranged in vertical columns with the digits of the same power placed under one another. These digits are then added to each other and if the total is greater than the base number (10 for denary and 2 for binary) then a carry is executed. The number carried is added to the next column and so on. In binary addition a carry occurs when the total of the two digits being added together is 2. In other words $1_2 + 1_2$ gives rise to a carry.

The following are the basic rules for binary addition.

$$0 + 0 = 0$$
$$0 + 1 = 1$$
$$1 + 0 = 1$$
$$1 + 1 = 0 \text{ carry } 1.$$

Table 1.7 shows a comparison between denary and binary addition of $823_{10} + 238_{10}$ and $11001_2 + 11011_2$.

Table 1.10 (a) Denary addition

	10^3 (1000)	10^2 (100)	10^1 (10)	10^0 (1)
		8	2	3
		2	3	8
Sum	1	0	6	1
Carry	1		1	

(b) Binary addition

	2^5 (32)	2^4 (16)	2^3 (8)	2^2 (4)	2^1 (2)	2^0 (1)
		1	1	0	0	1
		1	1	0	1	1
Sum	1	1	0	1	0	0
Carry	1	1		1	1	

Let us look at the binary addition in detail.

Units column: $1 + 1 = 0$, carry 1
2s column: $0 + 1 + \text{carry} = 1$, carry 1
4s column: $0 + 0 + \text{carry} = 1$
8s column: $1 + 1 = 0$, carry 1
16s column: $1 + 1 + \text{carry} = 1$, carry 1
32s column: carry $1 = 1$.

Where more than two binary numbers are added together, the carry may involve numbers other than 1. For instance while

$1 + 1 = 0$, carry 1 and
$1 + 1 + 1 = 1$, carry 1,

the following involve a carry greater than 1.

$$1 + 1 + 1 + 1 \quad = (1 + 1) + (1 + 1)$$
$$= (0, \text{carry } 1), + (0, \text{carry } 1)$$
$$= 0, \text{carry } 2;$$
$$1 + 1 + 1 + 1 + 1 = 1 + (1 + 1) + (1 + 1)$$
$$= 1, \text{carry } 2$$
$$0 + \text{carry } 2 \quad = 1, \text{carry } 1$$
$$1 + \text{carry } 2 \quad = 0, \text{carry } 2, \text{ and so on.}$$

EXAMPLE

Find the value of $10110_2 + 11101_2 + 11101_2$.

Solution
Arrange the binary numbers into columns:

	2^6	2^5	2^4	2^3	2^2	2^1	2^0
			1	0	1	1	0
			1	1	1	0	1
			1	1	1	0	1
Sum	1	0	1	0	0	0	0
Carry	1	2	2	2	1	1	

Binary subtraction

For the purposes of this section we will consider binary subtraction where a positive result is obtained. In this case the method employed is similar to that used in subtracting denary numbers. When a borrow is necessary in binary subtraction, a 1 is borrowed from the column immediately to the left, i.e. from the higher-order column.

The general rules for binary subtraction may be summarised as follows.

$$0 - 0 = 0$$
$$1 - 0 = 1$$
$$1 - 1 = 0$$
$$0 - 1 = 1 \text{ borrow } 1.$$

EXAMPLE

Subtract 0101_2 from 1111_2.

Solution

Arrange the two binary numbers into columns:

	2^3	2^2	2^1	2^0	
	(8)	(4)	(2)	(1)	
	1	1	1	1	
	0	1	0	1	
Result	1	0	1	0	(no borrow is involved).

In detail and starting from the least significant bit ($2^0 = 1$)

2^0 column $\quad 1 - 1 = 0$
2^1 column $\quad 1 - 0 = 1$
2^2 column $\quad 1 - 1 = 0$
2^3 column $\quad 1 - 0 = 1$

Answer: $1111_2 - 0101_2 = 1010_2$.

EXAMPLE

Subtract 1010_2 from 1100_2.

Solution

	2^3	2^2	2^1	2^0
	(8)	(4)	(2)	(1)
Borrow			→ (2^2)	
	1	1	0	0
	1	0	1	0
Result	0	0	1	0

In detail, and starting from the least significant bit,

2^0 column $\quad 0 - 0 = 0$
2^1 column $\quad 0 - 1 = 1$

In this case we need to borrow a 1 from the top number in the 2^2 column. Coming from the 2^2 column, the value of the borrow is twice that of the 2^1 column. Thus we get borrowed 1 (worth 2^2) $-$ 1 (worth 2^1) $=$ 1 (worth 2^1). **As a general rule borrowed 1 $-$ 1 $=$ 1.**

2^2 column $0 - 0 = 0$

The original 1 in the top number is now changed to 0 due to the borrow operation as shown by the arrow.

2^3 column $1 - 1 = 0$.

Answer: $1100_2 - 1010_2 = 0010$.

EXERCISE 1.5

Determine the value of the following binary operations.
(*a*) $11011 + 10110$
(*b*) $1011 + 1110 + 1101$
(*c*) $1110 - 101$
(*d*) $10111 - 1100$

Signed binary

So far we have considered positive or unsigned binary numbers only. For instance an eight-bit binary may have values between

$$0000\ 0000_2 = 00_{10} \text{ and } 1111\ 1111_2 = 255_{10},$$

all of which are positive. In order to distinguish a negative number from a positive one, a '$-$' sign is used to precede the denary, e.g. -25_{10}. In the binary system, the negative sign itself is given a binary code so that it may be recognised by a digital system. A binary bit known as the **sign bit** is introduced to the left of the MSB of the binary number which is devoted entirely to indicate the sign of the number. Such numbers are often referred to as sign-and-magnitude. When the sign bit is set to 0, the number is positive and when it is set to 1 the number is negative. It follows that a signed eight-bit number has its magnitude indicated by the first seven bits, 0–6, while its sign is indicated by bit 7 as shown in the following example.

Bit no.	7	6	5	4	3	2	1	0
	sign bit	2^6 (64)	2^5 (32)	2^4 (16)	2^3 (8)	2^2 (4)	2^1 (2)	2^0 (1)
	0	1	1	0	0	1	1	1

$$= + (64+32+4+2+1) = +103_{10}$$

	1	1	0	1	0	1	0	1

$$= - (64+16+4+1) = -85_{10}$$

	1	0	0	1	0	0	0	1

$$= - (16+1) = -17_{10}$$

Bit no.	7	6	5	4	3	2	1	0
	sign bit	2^6 (64)	2^5 (32)	2^4 (16)	2^3 (8)	2^2 (4)	2^1 (2)	2^0 (1)
	0	1	1	1	1	1	1	1

$= + (64+32+16+8+4+2+1) = +127_{10}$

	1	1	1	1	1	1	1	1

$= - (64+32+16+8+4+2+1) = -127_{10}$

	1	0	0	0	0	0	0	0

$= + 0 = 0$

	0	0	0	0	0	0	0	0

$= - 0 = 0$

Since only seven bits (0–6) can be used to indicate the magnitude, the maximum quantities that may be indicated by an eight-bit signed binary number are

$$[1] \quad 111\ 1111_2 = -127_{10},$$
and $\quad [0] \quad 111\ 1111_2 = +127_{10},$

with the bit in brackets as the sign bit.

In general an unsigned n-bit binary number can indicate a maximum magnitude of

$$M = 2^n - 1.$$

On the other hand, a signed n-bit binary number can indicate a maximum magnitude of

$$M = 2^{(n-1)} - 1.$$

Thus for an eight-bit register in a microprocessor system employing signed binary, the maximum magnitude that may be stored in the register is

$$\begin{aligned} M &= 2^{(n-1)} - 1 \\ &= 2^{(8-1)} - 1 \\ &= 2^7 - 1 \\ &= 128_{10} - 1 \\ &= 127_{10}, \end{aligned}$$

giving a range between -127_{10} and $+127_{10}$.

Ones and twos complement

The sign-and-magnitude system of representation is not conducive to arithmetic operations using positive and negative numbers. To overcome this limitation, the twos complement representation of negative numbers is used, which automatically produces the correct sign following arithmetic operations.

For positive binary numbers, the twos complement is identical to the sign-and-magnitude system in that it also uses the MSB as the sign bit, leaving the remaining bits to indicate the magnitude.

For negative numbers the twos complement is obtained by first producing the ones complement of the original binary number (with a positive sign) and then adding a 1 to the least significant bit. The complement of binary 1 is 0 and that of binary 0 is 1. Hence the complement of a binary number is obtained by inverting all its digits, i.e. changing all 1s to 0s and all 0s to 1s.

EXAMPLE

Find the twos complement of $+54$ and -54.

Solution

For a positive number, the twos complement is the same as the ordinary binary number. Hence

$$+54_{10} = [0]\ 0110110_2.$$

To find the twos complement for a negative number, start with its positive equivalent

	sign bit	
$+54$	= [0]	0110110
ones complement	[1]	1001001
		$+1$
twos complement	[1]	1001010
	= -54.	

Table 1.11 shows the conversion of denary numbers into ones and twos complement using eight-bit binary. It can be seen that in the

Table 1.11

Denary no.	Sign-and-magnitude			Ones complement			Twos complement		
$+127$	0	111	1111	0	111	1111	0	111	1111
$+126$	0	111	1110	0	111	1110	0	111	1110
$+2$	0	000	0010	0	000	0010	0	000	0010
$+1$	0	000	0001	0	000	0001	0	000	0001
Zero {	0	000	0000	0	000	0000	0	000	0000
	1	000	0000	1	111	1111	0	000	0000
-1	1	000	0001	1	111	1110	1	111	1111
-2	1	000	0010	1	111	1101	1	111	1110
-126	1	111	1110	1	000	0001	1	000	0010
-127	1	111	1111	1	000	0000	1	000	0001
-128	—			1	111	1111	1	000	0000

twos complement there is only one representation for zero, namely [0] 000 0000. As a result of this, the maximum negative magnitude has been increased by one to -128 giving a range of $+127$ and -128 for an eight-bit binary number.

Twos complement may be used to subtract two numbers. Consider $A - B$, where $A = 10101$ (21_{10}) and $B = 1100$ (12_{10}). Both A and B are positive and may be written as sign-and-magnitude numbers

$$A = [0]\ 10101$$
and $\quad B = [0]\ 01100.$

Note that in the case of B a 0 was added to the right of the sign bit to ensure that both numbers have the same number of bits. Taking the twos complement of B we get

$-B = [1]\ 10011 + 1 = [1]\ 10100.$
Since
$A - B = A + (-B),$
then
$A - B = A + \text{(twos complement of } B\text{)}$

A	[0] 10101
B's twos complement $\quad +$	[1] 10100
Sum	1 [0] 01001

The answer is either positive $(A > B)$ or negative $(A < B)$. This is indicated by the sign bit. When the result of the subtraction is positivie with a sign bit of [0], an extra bit is always introduced to its left as shown by the above example. This extra bit must be disregarded which, for the above example, produces a result of [0] 01001 $= +9_{10}$.

If on the other hand the result of the operation is negative with a sign bit of [1], then it must be converted back to a sign-and-magnitude number. To do this the same process of conversion to twos complement is carried out, i.e. inverting all the bits and adding a 1 to the LSB. For example to convert the twos complement number [1] 100 1010 ($= -54_{10}$)

[1]	1	0	0	1	0	1	0	twos complement number
[0]	0	1	1	0	1	0	1	after complementation
						$+$	1	add 1

[0] 0 1 1 0 1 1 0 $= 32 + 16 + 4 + 2 = 54$
in magnitude.

Exercise 1.6

Convert the following denary numbers to twos complement.
(a) 23 (b) -23 (c) -45 (d) -125

Multiplication

Binary multiplication has the following rules.

$$0 \times 0 = 0$$
$$0 \times 1 = 0$$
$$1 \times 0 = 0$$
$$1 \times 1 = 1$$

Binary multiplication may be performed following the same steps as long denary multiplication. For example to multiply $1110_2 = 14_{10}$ (known as the multiplicand) by $1101 = 13_{10}$ (known as the multiplier) the following procedure is followed.

	Binary				Denary	
Multiplicand	1	1	1	0	1	4
Multiplier	1	1	0	1	1	3
	1	1	1	0	4	2
	0 0 0 0				+1	4
	1 1 1 0					
	1 1 1 0					
Product	1 0 1 1 0 1 1 0				1 8 2	

Multiplication may also be performed by adding the multiplicand to itself a number of times equal to the multiplier. In the previous example, the same result may be obtained by adding 1110_2 to itself 1101_2 or thirteen times.

Division

Binary division may be carried out in the same manner as long denary division is carried out. For example to divide 110011 (known as the dividend) by 1001 (known as the divisor) the following procedure is followed.

```
Result                  1   0   1
                      ┌─────────────────
Divisor   1  0  0  1  │ 1   1   0   0   1   1
                      │−1   0   0   1
                      ├─────────────────
                      │ 0   0   1   1   1   1
                      │        −1   0   0   1
                      ├─────────────────
Remainder                         1   1   0
```

The result therefore is 101_2 with a remainder $= 110_2$.

Division may be also performed by repeated addition of the divisor until the dividend is reached or a remainder is obtained which is smaller than the dividend.

2 Logic gates

A logic gate is a two-state device, i.e. it has a two-state output: an output of zero volts representing logic 0 (or low) and a fixed voltage output representing logic of 1 (or high). The logic gate may have several inputs, all of which may be in one of the two possible logic states: 0 or 1. Logic gates may be used to perform several functions, e.g. AND, OR, NAND, NOR, NOT or EX-OR. Figure 2.1 shows the British and International standard symbols for these gates.

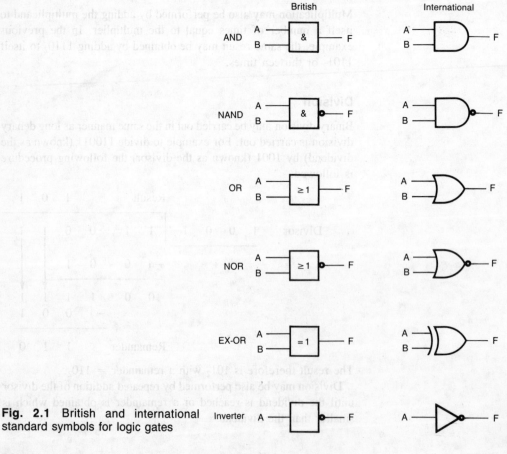

Fig. 2.1 British and international standard symbols for logic gates

AND gate

The function of the AND gate is to produce a logic 1 when all its inputs are at logic 1, otherwise a logic 0 will be obtained. The list of all possible combinations of the input and their respective outputs is known as the truth table of the gate. Table 2.1 shows the truth table of two-input gates.

A three-input AND gate is shown in Fig. 2.2 with its truth table shown in Table 2.2.

Table 2.1

Inputs		Output functions				
A	B	AND	NAND	OR	NOR	EX-OR
0	0	0	1	0	1	0
0	1	0	1	1	0	1
1	0	0	1	1	0	1
1	1	1	0	1	0	0

NAND gate

The NAND gate gives a logic 0 output when all its inputs are at logic 1. Conversely, if a logic 0 is present at any of the inputs of a NAND gate, its output will be at logic 1 (Table 2.1). NAND stands for NOT-AND, a function which is the opposite or the reverse of that of an AND gate.

OR gate

The OR gate gives an output of 1 if any of its inputs is at logic 1. Conversely, it requires a logic 0 at each of its inputs for a logic 0 to be obtained at the output (Table 2.1).

NOR gate

The NOR gate gives a logic 0 output if any of its inputs is at logic 1. Conversely, it requires a logic 0 at each of its inputs for a logic 1 to be obtained at the output. NOR stands for NOT-OR, a function which is the opposite of that of an OR gate.

NOT gate

The NOT gate is a single-input gate which functions as an inverter. When the input is high, the output is low and vice versa. The truth table is shown in Table 2.3.

International

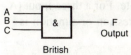

British

Fig. 2.2 Three-input AND gates

Table 2.2 Three-input AND gate truth table

Inputs			Output
A	B	C	F
0	0	0	0
0	0	1	0
0	1	0	0
0	1	1	0
1	0	0	0
1	0	1	0
1	1	0	0
1	1	1	1

Table 2.3 NOT gate truth table

Input A	Output F
0	1
1	0

MATTHEW BOULTON
COLLEGE LIBRARY

The EXCLUSIVE-OR gate

The EXCLUSIVE-OR (EX-OR) gate gives an output of 1 when either of its inputs is high but not when both are high or both are low. The truth table is shown in Table 2.1. It can be seen from the truth table that the EX-OR output is the binary sum of its two inputs.

Boolean expression

The function, i.e. the output, F of an individual gate or a combination of gates may be expressed in a logical statement known as a Boolean expression. This technique employs Boolean algebra with its own special notations and rules applicable to logic elements including gates.

Boolean algebra uses the following notations.

(*a*) The AND function is represented by a dot (.). Thus a two-input AND gate has an output or a function, F, represented by the Boolean expression

$$F = A.B \text{ or } B.A,$$

where A and B are the inputs to the gate. For a three-input (A, B, and C) AND gate the output expression or function is

$$F = A.B.C.$$

The dot is sometimes dispensed with, reducing the Boolean expression to AB and ABC, respectively.

(*b*) The OR function is represented by the plus symbol (+). Thus a two-input OR gate has an output function

$$F = A+B \text{ or } B+A.$$

(*c*) The NOT function is represented by a bar over the input. Thus a NOT gate with an input A has an output function

$$F = \overline{A}$$

(read as not A).

(*d*) The EXCLUSIVE-OR function is given the special symbol $\oplus$. For a two-input EX-OR gate the output function will therefore be

$$F = A \oplus B.$$

The NOT notation may be used to represent any inverting function. For example, if the output of an AND gate is inverted to produce the NAND function, the Boolean expression is given as

$$F = \overline{A.B} \text{ (or } \overline{AB}).$$

Similarly for the NOR function, the Boolean expression is

$$F = \overline{A+B}.$$

Table 2.4 gives a summary of the basic Boolean notations.

Table 2.4 Boolean notation

Function	Boolean notation
AND	A . B
OR	A + B
NOT	$\overline{A}$
EX-OR	A $\oplus$ B
NAND	$\overline{A . B}$
NOR	$\overline{A + B}$

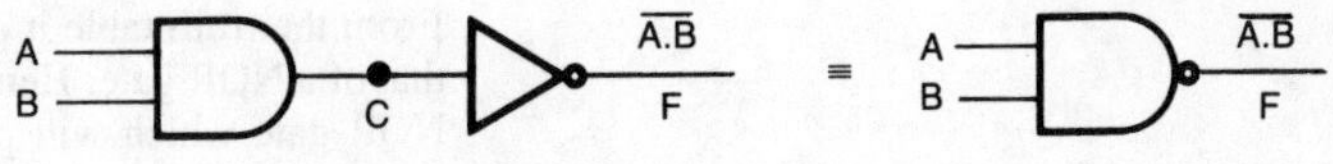

Fig. 2.3

Combinational gates

Logic systems usually involve more than one gate, forming a combination of gates to perform a certain function. For instance, the simple combination of an AND gate and a NOT gate shown in Fig. 2.3(a) makes up the NAND gate shown in (b). Using Boolean expressions

output of AND gate, $C = A.B$
output of NOT gate, $F = \overline{A.B}$.

Similarly, a NOR function may be constructed by an OR gate followed by a NOT gate as shown in Fig. 2.4. The Boolean expressions are

output of OR gate, $C = A+B$
output of NOT gate, $F = \overline{A+B}$.

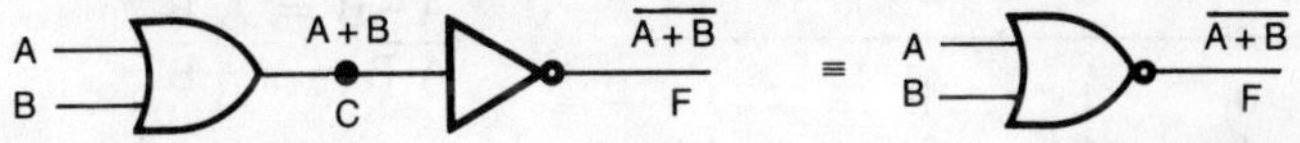

Fig. 2.4

EXAMPLE

Refer to Fig. 2.5.

(a) Find the Boolean expression of the function of the combination of gates.
(b) Construct the truth table showing the logic states at all points and hence prove that the combination may be reduced to a single gate.
(c) Prove that $\overline{A+B} = \overline{A}.\overline{B}$

Solution

(a) Boolean expression at $C = \overline{A}$.
  Boolean expression at $D = \overline{B}$.
  Boolean expression at $F = \overline{A}.\overline{B}$.
(b) The truth table is:

Fig. 2.5

Inputs				Output
A	B	C	D	F
0	0	1	1	1
0	1	1	0	0
1	0	0	1	0
1	1	0	0	0

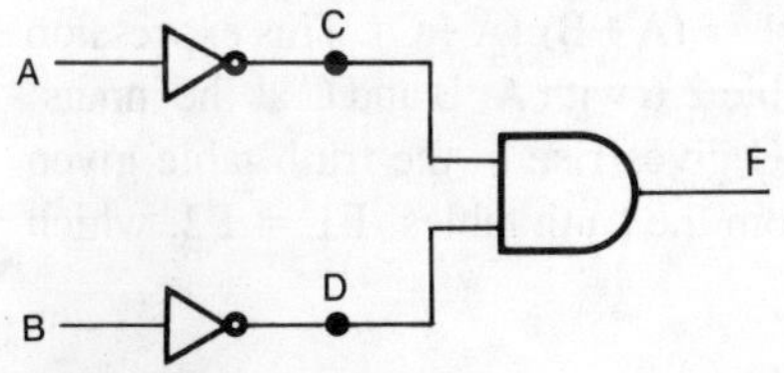

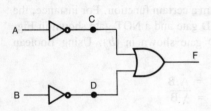

Fig. 2.6

Table 2.5

| Inputs | | | | Output |
A	B	C	D	F
0	0	1	1	1
0	1	1	0	1
1	0	0	1	1
1	1	0	0	0

From the truth table it can be seen that the output at F is identical to that of a NOR gate. Hence the combination may be reduced to a single NOR gate which will produce the same output.

(c) The NOR function has a Boolean expression of $\overline{A+B}$. However the expression produced in (a) above is $\overline{A}.\overline{B}$. It follows therefore that $\overline{A+B} = \overline{A}.\overline{B}$.

From the above it can be deduced that 'not A or B' is equivalent to 'not A *and* not B'. This is known as the first De Morgan theorem.

The second De Morgan theorem may be deduced from the logic circuit in Fig. 2.6.

Boolean expression at $C = \overline{A}$.
Boolean expression at $D = \overline{B}$.
Boolean expression at $F = \overline{A}+\overline{B}$.

However, from the truth table shown in Table 2.5 it can be seen that the output is identical to that of a single NAND gate. Since the NAND function is $\overline{A.B}$, then

$$\overline{A.B} = \overline{A}+\overline{B}.$$

From the two De Morgan theorems

$$\overline{A+B} = \overline{A}.\overline{B} \tag{1}$$
$$\overline{A.B} = \overline{A}+\overline{B}. \tag{2}$$

It follows that the complement of a function may be obtained by complementing each variable and changing the (.) to a (+) and the (+) to a (.). For example, the complement of A+B is $\overline{A+B}$ which may be written as $\overline{A}.\overline{B}$. Similarly, the complement of A.B is $\overline{A.B} = \overline{A}+\overline{B}$.

Other Boolean theorems are:

$$A.B = B.A$$
$$A+B = B+A$$
$$A.(B.C) = (A.B).C$$
$$A+(B+C) = (A+B)+C$$
$$(A+B).(A+C) = A+B.C$$
$$A.B+A.C = A.(B+C)$$
$$A+A.B = A$$
$$A.(A+B) = A$$
$$A+\overline{A}.B = A+B$$
$$A.(\overline{A}+B) = A.B$$

The above theorems may be proved by constructing the truth table in each case. For example to prove that $(A+B).(A+C) = A+B.C$, begin with the first expression $F1 = (A+B).(A+C)$. This expression gives the truth table shown in Table 2.6 with A, B and C as the inputs.

The expression $F2 = A+B.C$ gives rise to the truth table given in Table 2.7. As can be seen from the truth tables, $F1 = F2$, which proves that

$$(A+B).(A+C) = A+B.C.$$

Table 2.6

A	B	C	A + B	A + C	F1 = (A + B) . (A + C)
0	0	0	0	0	0
0	0	1	0	1	0
0	1	0	1	0	0
0	1	1	1	1	1
1	0	0	1	1	1
1	0	1	1	1	1
1	1	0	1	1	1
1	1	1	1	1	1

Table 2.7

A	B	C	B.C	F2 = A + B.C
0	0	0	0	0
0	0	1	0	0
0	1	0	0	0
0	1	1	1	1
1	0	0	0	1
1	0	1	0	1
1	1	0	0	1
1	1	1	1	1

EXAMPLE

Prove that

$$A.(\overline{A}+B) = A.B.$$

Solution

The truth tables for $F1 = A.(\overline{A}+B)$ and $F2 = A.B$ are shown in Table 2.8. As can be seen, functions F1 and F2 are identical and are therefore equal.

Boolean rules and theorems may be used to simplify the design of combinational logic circuits without having to draw lengthy truth tables.

Table 2.8

A	B	$\overline{A}$	$\overline{A}+B$	F1 = A.($\overline{A}$ + B)	F2 = A.B
0	0	1	1	0	0
0	1	1	1	0	0
1	0	0	0	0	0
1	1	0	1	1	1

EXAMPLE

Using Boolean algebra, simplify the expression

$$F = \overline{\overline{(A.B)}A.\overline{(A.B)}B}.$$

Solution

The complement of F is

$$\overline{F} = \overline{(A.B)}A.\overline{(A.B)}B.$$

Taking the complement again and using De Morgan's theorem number 1 we get

$$F = \overline{(A.B)}A + \overline{(A.B)}B.$$

Using De Morgan's theorem number 2 for A.B, we get

$$F = (\overline{A}+\overline{B})A + (\overline{A}+\overline{B})B.$$

Opening the bracket

$$F = A\bar{A} + A\bar{B} + \bar{A}B + B\bar{B}.$$

But

$$A.\bar{A} = 0 \text{ and } B.\bar{B} = 0,$$

then

$$F = A.\bar{B} + \bar{A}.B.$$

In other words, a logic 1 is obtained when either A = 1 and B = 0 OR A = 0 and B = 1. This of course is the EXCLUSIVE-OR function. It follows therefore that

$$F = A \oplus B.$$

Figure 2.7 shows the circuit for the original expression and its simplified form.

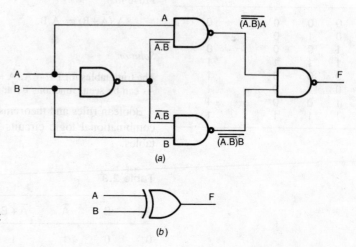

Fig. 2.7   (a) Combinational logic circuit and (b) its equivalent

EXERCISE 2.1

Refer to Fig. 2.8.

(a) Construct a truth table for the combinational logic circuit and hence state (i) the Boolean expression of the output and (ii) its function.
(b) If gate G1 is faulty, with its output P permanently stuck at logic 1, state which input combination or combinations will give false output at S.

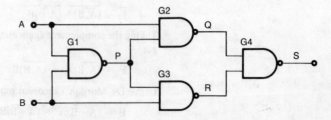

Fig. 2.8

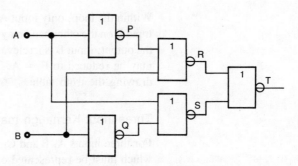

Fig. 2.9

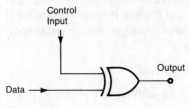

Fig. 2.10

EXERCISE 2.2

Refer to Fig. 2.9.

(*a*) State the Boolean expression at each of the nodes P, Q, R, S, and T.
(*b*) Construct a truth table for the combinational logic circuit and hence show how it can be reduced to a combination of two gates only.

EXERCISE 2.3

State the effect on the data fed into the gate in Fig. 2.10 when the control input is (*a*) at logic 0, and (*b*) at logic 1.

Karnaugh map

As it has been shown, Boolean algebra may be used to simplify the design of logic circuits. However this method involves lengthy mathematical operations. An alternative method is the use of the Karnaugh map (K-map) or diagrams based on pattern-recognition techniques.

The Karnaugh map contains all possible combinations of the binary inputs of a logic system. These combinations are contained in squares arranged in rows and columns. The simplest map is that for two inputs or variables, A and B shown in Fig. 2.11. The columns represent input A, with the left-hand column representing A = 0 and the right-hand column representing A = 1. On the other hand, the rows represent input B with the top row representing B = 0 and the bottom row representing B = 1. The four boxes thus obtained represent all $2^2 = 4$ possible combinations of the inputs. For a given Boolean expression, the output of any combination of inputs is written inside the appropriate box. For example consider the expression

$$F = A.\overline{B} + A.B.$$

The Karnaugh map will be as shown in Fig. 2.12. A logic 1 is placed inside the square corresponding to the first part of the expression, namely $A.\overline{B}$ (A.not B), i.e. a 1 is placed in the square A = 1, B = 0. This is repeated for the second part of the expression, namely A.B, at A = 1, B = 1. The remaining squares may remain blank or have 0 placed in them. The adjacent 1s are looped as shown in Fig. 2.13.

Fig. 2.11 Two-variable Kanaugh map

Fig. 2.12

Fig. 2.13

Within the loop, only input A is not changing (it remains at logic 1). Input B on the other hand may take the value of 1 or 0 without affecting the output; input B is irrelevant. It follows therefore that the function may be reduced to F = A. This simplification may be verified by drawing the truth table.

Three-input Karnaugh map

For three inputs A, B and C, there are $2^3 = 8$ different combinations which must be represented on a K-map as shown in Fig. 2.14. The combinations of A and B are not arbitrarily assigned along the columns. They are arranged as an orderly progression in which the value of only one input changes from square to square. This also applies to the edges. Thus, the difference between the right edge and the left edge is A changing from 1 to 0 while B remains constant at 0. Similar reasoning applies to the other two edges, the top and bottom. Consider the Boolean expression

$$F = \overline{A}B\overline{C} + \overline{A}BC = A\overline{B}\overline{C} + A\overline{B}C.$$

The K-map is as shown in Fig. 2.15 with logic 1 placed in boxes corresponding to the following:

$\overline{A}B\overline{C}$ (notA.B.notC) at A = 0, B = 1, C = 0,
$\overline{A}BC$ (notA.B.C) at   A = 0, B = 1, C = 1,
$A\overline{B}\overline{C}$ (A.notB.notC) at A = 1, B = 0, C = 0,
$A\overline{B}C$ (A.notB.C) at   A = 1, B = 0, C = 1.

The remaining boxes are left blank for neatness. Two loops, grouping adjacent 1s together may be identified. The loop to the left represents A = 0, B = 1, i.e. $\overline{A}B$. The loop to the right represents A = 1, B = 0, i.e. $A\overline{B}$. In both cases the value of input C is irrelevant. In other words, the output is not a function of C. An output may therefore be produced with either $\overline{A}B$ OR $A\overline{B}$, giving the simplified expression

$$F = \overline{A}B + A\overline{B}.$$

EXAMPLE

Simplify the Boolean expression

$$F = \overline{A}B\overline{C} + \overline{A}BC + AB\overline{C} + ABC.$$

Solution

The K-map is shown in Fig. 2.16(*a*) with logic 1 entered in the boxes

$\overline{A}B\overline{C}$, $\overline{A}BC$, $AB\overline{C}$ and ABC.

Two loops may be identified giving a Boolean expression of

$$F = \overline{A}B + AB.$$

However, a further simplification may be obtained by a larger loop containing *all* adjacent logic 1s as shown in Fig. 2.16(*b*). Both A and C change value as you cross from one square to the other with only input B remaining constant.

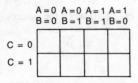

Fig. 2.14 Three-variable Karnaugh map

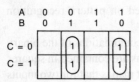

Fig. 2.15

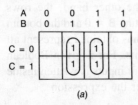

(a)

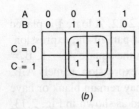

(b)

Fig. 2.16

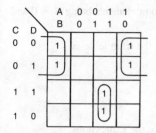

Fig. 2.17

Table 2.9

A	B	C	F
0	0	0	1
0	0	1	1
0	1	0	1
0	1	1	1
1	0	0	0
1	0	1	1
1	1	0	1
1	1	1	0

The values of A and C are therefore irrelevant to the output which is a function of B only. This gives the simplified function of

$$F = B.$$

Loops may also be drawn from one edge to the other. For example consider the four-variable expression

$$F = \overline{A}\overline{B}\overline{C}\overline{D} + A\overline{B}\overline{C}\overline{D} + \overline{A}\overline{B}C\overline{D} + A\overline{B}C\overline{D} + ABCD + ABC\overline{D}.$$

The K-map is shown in Fig. 2.17. A logic 1 is placed at

$$\overline{A}\overline{B}\overline{C}\overline{D},\ A\overline{B}\overline{C}\overline{D},\ \overline{A}\overline{B}C\overline{D},\ A\overline{B}C\overline{D},\ ABCD\ \text{and}\ ABC\overline{D}.$$

Two loops can be identified. The two half loops at the top left- and right-hand sides form one loop with B and C remaining constant at logic 0 throughout giving the expression $\overline{B}\overline{C}$. The second loop has A, B and C remaining constant at logic 1 giving the expression ABC. The simplified expression therefore is

$$F = \overline{B}\overline{C} + ABC.$$

EXAMPLE

Construct a simplified Boolean expression that will perform the function described by the truth table in Table 2.9.

Solution

The first step is to write the Boolean expression for each combination of inputs that produces an output of logic 1 (Table 2.10). This gives a Boolean expression for the output of

$$\overline{A}.\overline{B}.\overline{C} + \overline{A}.\overline{B}.C + \overline{A}.B.\overline{C} + \overline{A}.B.C + A.\overline{B}.C + A.B.\overline{C}.$$

Table 2.10

A	B	C	F	Boolean expression
0	0	0	1	$\overline{A}.\overline{B}.\overline{C}$
0	0	1	1	$\overline{A}.\overline{B}.C$
0	1	0	1	$\overline{A}.B.\overline{C}$
0	1	1	1	$\overline{A}.B.C$
1	0	0	0	
1	0	1	1	$A.\overline{B}.C$
1	1	0	1	$A.B.\overline{C}$
1	1	1	0	

Using this expression, the Karnaugh map can be drawn as shown in Fig. 2.18. Three loops may be drawn as shown. Where an overlap is necessary for all the 1s to be included as in the case of loops 2 and 3, the overlap is kept to a minimum, i.e. to a single square in each case. The following loops are then identified:

(*a*) loop 1 (unbroken line) with the only variable common to all squares being A = 0 or $\overline{A}$;
(*b*) loop 2 (broken line) with the common variables being B = 1 and C = 0 or $B.\overline{C}$ and

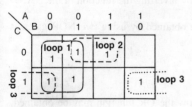

Fig. 2.18

(c) loop 3 (dotted line) with the common elements being B = 0 and C = 1 or $\overline{B}.C$.

It follows that the simplified expression for the truth table is

$$\overline{A} + B.\overline{C} + \overline{B}.C.$$

EXERCISE 2.4

Using the Karnaugh map technique, simplify the expressions:

(a) $A.\overline{B} + A.B + \overline{A}.B$;
(b) $X.Y.Z + X.\overline{Y}.Z + X.\overline{Y}.\overline{Z} + X.Y.\overline{Z}$;
(c) $A.\overline{C} + \overline{A}.(B+C) + A.B.(C+\overline{B})$.

Gate conversion

One major consideration in designing a logic circuit (other than the number of logic elements) is the type of logic gate employed. Logic gates are manufactured in integrated circuit (IC) packages containing a number of identical gates, e.g. a quad two-input NAND (containing four two-input NAND gates), a dual four-input AND (containing two four-input AND gates) and a quad two-input OR (containing four two-input OR gates). Higher levels of integration produce IC packages containing a much larger number of gates.

In logic design discrete logic gates are not normally used. Instead logic IC packages are used with multiple gates. It is therefore desirable, if only to reduce the number of ICs used, that one type of gate is employed, and hence one type of IC, rather than a mix of different gates. For this reason gate conversion is employed in which a particular gate function is implemented by using a combination of other gates all of the same type.

The NOT function for instance may be obtained from a NAND gate with shorted inputs as shown in Fig. 2.19(a). Similarly, a shorted-inputs NOR gate acts as a NOT gate (Fig. 2.19(b)).

EXAMPLE

Using NAND gates only, show how the function of an AND gate may be obtained.

Solution

The AND function may be obtained by inverting the function of the NAND gate as shown in Fig. 2.20.

Similarly the OR function may be obtained by using two NOR gates as shown in Fig. 2.21.

EXAMPLE

By using NAND gates only show how the OR function may be obtained.

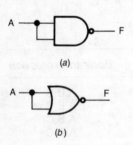

(a)

(b)

Fig. 2.19 NOT function

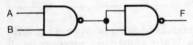

Fig. 2.20 AND function

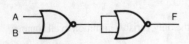

Fig. 2.21 OR function

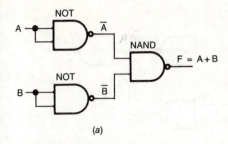

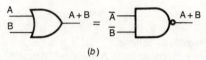

Fig. 2.22 OR function

Solution

The function of an OR gate is given by

$$F = A + B.$$

By using De Morgan's theorem number 1 we get

$$\overline{A+B} = \overline{A}.\overline{B}.$$

By complementing each side we get

$$A+B = \overline{\overline{A}.\overline{B}}.$$

This gives the logic circuit shown in Fig. 2.22(*a*).

This is a very useful conversion rule. It states that an OR gate may be replaced by a NAND gate if each of the inputs are complemented as shown in Fig. 22.2(*b*).

EXAMPLE

By using NOR gates only, show how the AND function may be produced.

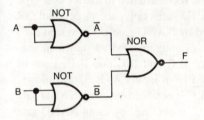

Fig. 2.23 AND function

Solution

The AND function is given by

$$F = A.B.$$

Using De Morgan's theorem number 2 we get

$$\overline{F} = \overline{A.B} = \overline{A}+\overline{B}$$
$$\therefore \; F = \overline{\overline{A}+\overline{B}}.$$

This gives the logic circuit in Fig. 2.23.

EXAMPLE

Construct a logic circuit to perform the function described in the truth table in Table 2.11.

Solution

The first step is to construct a simplified Boolean function for the truth table (see EXAMPLE on p. 25), namely

$$F = \overline{A}+B.\overline{C} + \overline{B}.C.$$

The simplified function may be produced by feeding each of its component parts, $\overline{A}$, $B.\overline{C}$ and $\overline{B}.C$ into a three-input OR gate. G6, as shown in Fig. 2.24.

Table 2.11

A	B	C	F
0	0	0	1
0	0	1	1
0	1	0	1
0	1	1	1
1	0	0	0
1	0	1	1
1	1	0	1
1	1	1	0

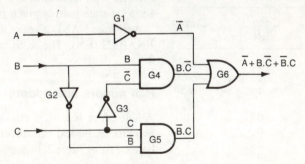

Fig. 2.24

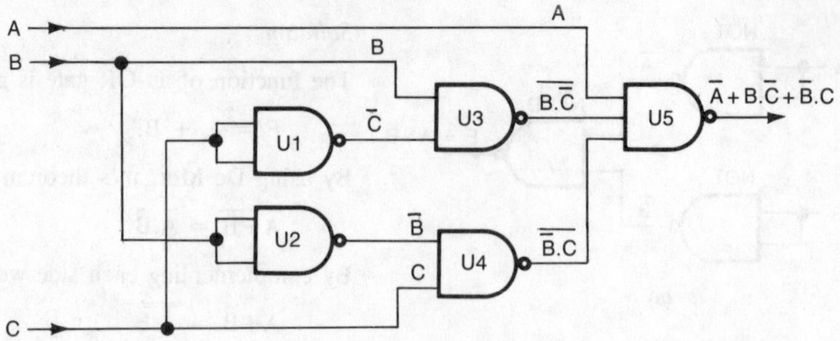

Fig. 2.25

The inputs to the OR gate are obtained, as follows.

The OR gate input $\bar{A}$ is obtained by inverting input A using inverter gate G1, $B.\bar{C}$ is obtained by using AND gate G4 and inverter G3, and $\bar{B}.C$ is obtained by using AND gate G5 and inverter G2.

The logic circuit in Fig. 2.24 uses three types of logic gates hence three different IC packages, making it impractical and uneconomical. To avoid this, the circuit should be reconstructed to use a single type of gate. For example, using NAND gates only, the three-input OR gate in Fig. 2.24 may be replaced by NAND gate U5 provided each component part of the function is complemented as shown in Fig. 2.25, namely $\bar{A}$ into A, $B.\bar{C}$ into $\overline{B.\bar{C}}$ and $\bar{B}.C$ into $\overline{\bar{B}.C}$. Each of these component parts is obtained from inputs A, B, C as follows. Input A is fed directly into U5, $\overline{B.\bar{C}}$ is obtained by using NAND gate U3 and inverter gate U1, and $\overline{\bar{B}.C}$ is obtained by using NAND gate U4 and inverter gate U2.

The tri-state gate

In many applications such as data transfer in microprocessor systems, a third state, the high-impedance (or open-circuit) state, is necessary. Such gates are known as tri-state, employing an additional control input. The output of a tri-state gate is valid only when the control input is ENABLED. The control input is usually active low, i.e. the gate is active when the control goes to logic 0. When the control input is NOT ENABLED, the output of the gate is forced into high impedance, Z or open circuit (o/c). Figure 2.26 shows a tri-state inverter with its control line active low $(\overline{ENABLE})$. The truth table for a tri-state gate is given in Table 2.12.

A tri-state NAND gate is shown in Fig. 2.27 with active low control, $\overline{ENABLE}$ (EN). The truth table is as shown in Table 2.13.

Half adders and adders

A half adder is a logic circuit whose output function is the sum of *two* bits. A typical half-adder circuit is shown in Fig. 2.28. As can be seen from Table 2.14, output S represents the sum of the two input bits A and B with output C indicating a carry bit if any.

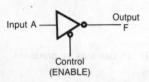

Fig. 2.26 Tri-state inverter

Table 2.12

Input A	Control $\overline{EN}$	Output F
0	1	o/c
1	1	o/c
0	0	1
1	0	0

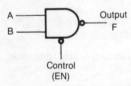

Fig. 2.27 Tri-state NAND gate

Table 2.13

Inputs A	B	Control $\overline{EN}$	Output F
0	0	1	o/c
0	1	1	o/c
1	0	1	o/c
1	1	1	o/c
0	0	0	1
0	1	0	1
1	0	0	1
1	1	0	0

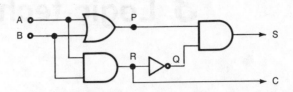

Fig. 2.28 Half adder

Table 2.14

A	B	P	R	Q	S	C
0	0	0	0	1	0	0
0	1	1	0	1	1	0
1	0	1	0	1	1	0
1	1	1	1	0	0	1

A half-adder may also be constructed using NAND gates only as shown in Fig. 2.29. From this truth table it can be seen that the sum of the two bits is their EXCLUSIVE−OR function giving

$$S = \overline{A}.B + A.\overline{B} = A \oplus B$$

The carry bit is given by $C = A.B$.

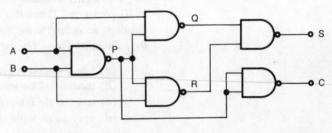

Fig. 2.29 Half adder using NAND gates

If the bits to be added include a carry bit, C_{in}, obtained from a previous calculation, then a full adder has to be used. A full adder, normally referred to as an adder consists of two half-adders as shown in Fig. 2.30. The truth table is as shown in Table 2.15.

From the truth table it can be deduced that:

$$S_{out} = \overline{C}_{in}\overline{A}B + C_{in}\overline{A}\overline{B} + \overline{C}_{in}A\overline{B} + C_{in}AB \text{ and}$$
$$C_{out} = C_{in}A + C_{in}B + AB.$$

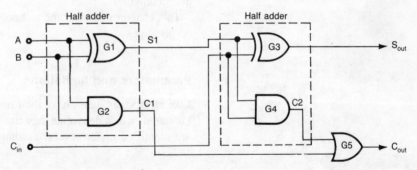

Fig. 2.30 Full adder

Table 2.15

A	B	C_{in}	S1	C1	C2	S_{out}	C_{out}
0	0	0	0	0	0	0	0
0	0	1	0	0	0	1	0
0	1	0	1	0	0	1	0
0	1	1	1	0	1	0	1
1	0	0	1	0	0	1	0
1	0	1	1	0	1	0	1
1	1	0	0	1	0	0	1
1	1	1	0	1	0	1	1

MATTHEW BOULTON
COLLEGE LIBRARY

3 Logic technology

Logic elements including gates and memory devices are manufactured in IC packages. These ICs are classified into categories, known as families, according to the number of gates or equivalent elements that they contain. These families are:

(a) small-scale integration (SSI), up to 10 gates;
(b) medium-scale integration (MSI), 10−100 gates;
(c) large-scale integration (LSI) 100−1000 gates;
(d) very large-scale integration (VLSI) 1000−10 000; and
(e) super large-scale integration (SLSI) 10 000−100 000.

The level of integration indicates the complexity of the IC package. It increases in powers of 10, i.e. 10, 100, 1000 and so on. Small- and medium-scale integration (SSI and MSI) provide discrete logic elements such as gates, counters and registers. Large- and very large-scale integration (LSI and VLSI) provide memory chips, microprocessors and complete systems such as complete single-chip microcomputers.

This chapter describes the technologies employed in SSI and MSI IC packages.

Parameters and limitations

Like all devices, gates have their own parameters or properties and limitations which determine their use and application. These properties are identified under several headings.

1. **Power dissipation** is the power consumed by the gate when fully driven by all its inputs.
2. **Fan-in** is the number of similar logic gates that can be connected to the input without any degradation of the voltage levels.
3. **Fan-out** is the number of similar logic gates that a gate is capable of driving without any degradation of the voltage levels.
4. **Speed** is expressed in terms of the delay time experienced by an input pulse. This delay is often referred to as propagation delay. **Propagation delay** is the time that elapses between the

application of a pulse edge at the input of a logic device and the consequential change in the logic state of the output.

5. **Noise immunity** is the maximum noise voltage that may appear at the input of a logic element without causing its output to change. Noise is a general term used to refer to all unwanted signals, e.g. hum, transients and glitches. There are two types of noise immunity: the **low-noise immunity** (LNI) when the input is at logic 0, and the **high-noise immunity** (HNI) when the input is at logic 1. The latter, i.e. a logic 1 input, can normally withstand a larger level of noise voltage than can the former. A **noise margin** is usually quoted by the manufacturers. This refers to the amplitude of a noise pulse that may cause the logic level to change.

Logic technology

Logic elements are broadly divided into two categories: those using bipolar semiconductors and those using metal-oxide silicon (MOS) technologies. The bipolar categories include the resistor−transistor logic (RTL) and the transistor−transistor logic (TTL). The MOS type uses field-effect transistors and includes p-channel (pMOS), n-channel (nMOS) and complementary (CMOS).

Logic threshold

A logic element has two separate states; logic 0 represented by low voltage, normally 0 V and logic 1 represented by a positive voltage for positive logic (a negative voltage is used for negative logic). The voltage level which represents logic 1 depends on the technology used. For bipolar technology, logic 1 is represented by 5 V, while for MOS, logic 1 may vary between 3 V to over 15 V. In practice however, there are two threshold voltages, one for each logic level. The logic 1 threshold is the voltage above which a logic 1 is recognised and the logic 0 threshold is the voltage below which a logic 0 is recognised. The precise logic threshold voltages to which logic elements will respond are for bipolar technology (e.g. TTL),

logic 1 threshold = 2 V (or 2.4 V with a noise margin of 0.4 V)

logic 0 threshold = 0.8 V (or 0.4 V with a noise margin of 0.4 V),

and, for MOS technology (e.g. CMOS),

logic 1 threshold = $0.7 \times$ supply voltage V_{DD}
logic 0 threshold = $0.3 \times$ supply voltage V_{DD}.

A logic level between the two stated threshold voltages (bipolar $0.8-2$ V; MOS $0.3V_{DD}-0.7V_{DD}$) is indeterminate, neither logic 1 nor logic 0.

Bipolar logic

Bipolar logic uses bipolar transistors to produce the various logic functions. A bipolar transistor may be used in two ways; the saturation mode and the non-saturation mode.

In the saturation mode, the transistor is used as a switch. When the base−emitter (b−e) junction is forward biased, collector current flows as shown in Fig. 3.1(a). If the parameters of the transistor and the values of R_1 and R_2 are correctly chosen, the transistor may be made to saturate with collector current at maximum value. At saturation, the emitter−collector potential $V_{CE(sat)}$ is in the region of 0.1−0.2 V with the transistor acting as a virtual short circuit. When the b−e junction is reverse biased (Fig. 3.1(b)) the transistor is turned off with its collector at supply voltage V_{CC} (5 V).

The advantage of the saturation mode of operation is its low power consumption. It does, however, suffer from long propagation delay. Figure 3.2 shows the response of the collector output voltage V_{CE} of a transistor to an ideal input pulse waveform. As can be seen, the collector potential does not immediately respond to the input waveform, instead a time or propagation delay is introduced.

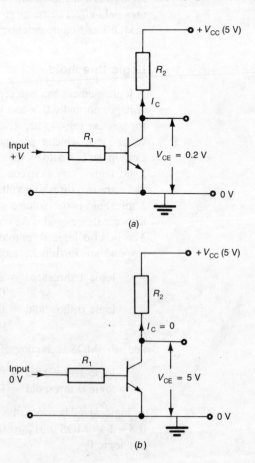

Fig. 3.1 (a) saturated transistor, and (b) cut-off transistor

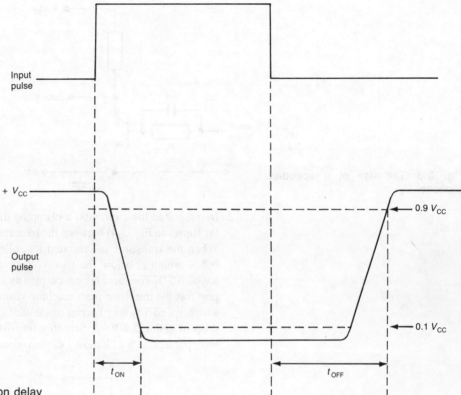

Fig. 3.2 Propagation delay

Types of propagation delay

There are two different types of propagation delay associated with the transistor: a low-to-high propagation delay t_{ON} introduced when a positive step input turns a cut-off transistor ON giving a negative step output as shown in Fig. 3.2. The other type of propagation delay is the high-to-low propagation delay t_{OFF} introduced when a saturated transistor is turned OFF. It should be noted that t_{ON} is larger than t_{OFF} since a saturated transistor takes longer to reach cut-off than a cut-off transistor takes to saturate.

Each propagation delay t_{ON} and t_{OFF} consists of two distinct parts. It is usual for a single value of propagation delay t_{PD} to be given which is the average of t_{ON} and t_{OFF}

$$t_{PD} = \tfrac{1}{2}(t_{ON} + t_{OFF}).$$

The typical propagation delay of a bipolar transistor gate is of the order of $t_{ON} = 7$ ns and $t_{OFF} = 11$ ns. Such a delay results in relatively slow switching speeds. There are several ways of improving the switching speed of a transistor.

1. A speed-up capacitor may be connected across the base resistor (Fig. 3.3). This method is popular in discrete logic circuits.

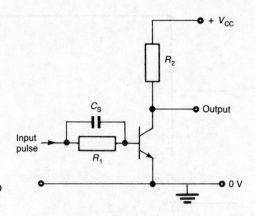

Fig. 3.3 The use of a speed-up capacitor C_s

2. In integrated logic circuits, a clamping diode D_1 is connected (as shown in Fig. 3.4) between the base and collector terminals. When the transistor is saturated its collector is at $V_{CE(sat)} = 0.2$ V which is below the base voltage of 0.7 V. When that happens, D_1 conducts taking current away from the base and prevents the transistor from reaching saturation. D_1 is normally a high-speed Schottky barrier diode with a low forward voltage drop of about 0.4 V. A Schottky transistor (Fig. 3.5) which incorporates such a clamping diode within it may also be used.

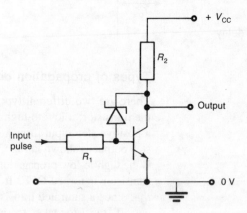

Fig. 3.4 The use of Schottky diode

Fig. 3.5 Schottky transistor

3. Propagation delay may be improved by minimising the effect of stray capacitance, C_s, which includes inter-electrode and junction capacitances as well as the input capacitance of the next stage. The effect of C_s may be reduced by improved layout and miniaturisation, hence the desirability of IC gate packages. The effect of stray capacitance may be further reduced by the use of the **totem pole arrangement**. Stray capacitance, C_s, is effectively connected across the output as shown in Fig. 3.6. When the transistor is turned on, C_s is quickly discharged through the transistor giving a short time constant and hence an improved delay time. However, when the transistor is turned

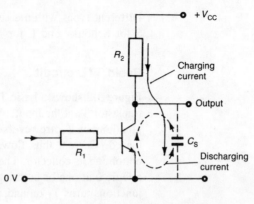

Fig. 3.6 The effect of stray capacitance, C_s

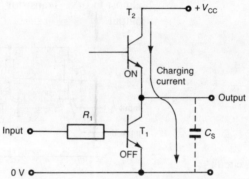

Fig. 3.7 Totem pole arrangement

off, C_s charges through load resistor R_2, giving a relatively long rise-time. The time constant $C_s R_2$ may be reduced by decreasing the value of R_2. However this increases power dissipation of the transistor. To avoid this increase in power consumption, a dynamic load resistor may be used in the form of a transistor in what is known as the totem pole arrangement. R_2 is replaced by a second transistor T_2 as shown in Fig. 3.7. T_2 is made to turn on when T_1 is off. Stray capacitance C_s then charges up through the low resistance presented by on transistor T_2, thus reducing rise-time. T_2 turns off when T_1 is on, maintaining the small discharge time constant through T_1.

Integrated circuit logic

There are several ways of building a gate circuit such as the resistor—transistor logic (RTL) and the diode—transistor logic (DTL). Modern gates are manufactured in integrated circuit form and use such technologies as transistor—transistor logic (TTL) and the complementary metal-oxide silicon (CMOS). Each type is given a series or family prefix number which is used by all IC manufacturers. For example, the standard TTL is given the prefix number of 74 while a CMOS has a prefix of 40, for example, the TTL 7400 (4 × two-input NAND gates) and the CMOS 4001 (4 × two-input NOR gates).

Different types within a series are identified by a letter code, e.g. S for Schottky and L for low-power.

Basic TTL circuit

Figure 3.8 shows a basic TTL NAND gate using a multiple-emitter transistor T_1 at the input. When all inputs are high (+5 V), the b—e junctions of T_1 are reverse biased but its b—c junction is forward biased. Current thus flows into the base of T_2 turning it on and bottoming its collector. The output is thus low (logic 0). Conversely, a logic 0 at one or both inputs, forward biases one or both T_1 b—e junctions turns T_1 on and forces its collector (and T_2 base) to drop almost to 0 V. Transistor T_2 thus turns off to give a high (logic 1) output.

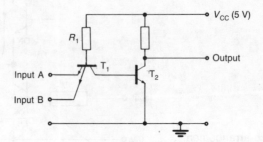

Fig. 3.8   Basic TTL circuit

Standard TTL circuit

The basic TTL gate is not used by manufacturers as it suffers from low operating speed and a small fan-out capability. IC manufacturers use the standard (also known as the totem-pole) TTL NAND gate shown in Fig. 3.9 which employs an output stage in addition to the basic circuit. As mentioned earlier, the totem pole arrangement reduces propagation delay thus increasing its operating speed.

Referring to Fig. 3.9, T_1 is a multi-emitter transistor which turns

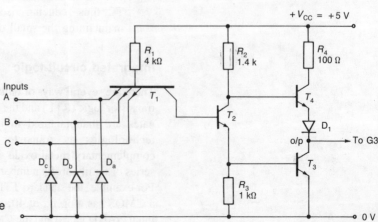

Fig. 3.9   TTL NAND gate totem pole circuit

T_2 on when all inputs are high. Otherwise T_2 is off. When T_2 is off, its collector is at V_{CC} (turning T_4 on) and its emitter is at 0 V potential (turning T_3 off). Conversely when T_2 is on, its collector is low, at about 0.8 V (turning T_4 off), and its emitter is at a potential of about 0.7 V (turning T_3 on). When T_4 is on, T_3 is off and vice versa.

The two transistors T_3 and T_4 thus function as a potential divider connected across the supply voltage V_{CC}. When T_4 is on, it represents a low impedance or short circuit, while T_3 being off represents a high impedance or open circuit (Fig. 3.10(a)). The output is therefore high (logic 1). Conversely, when T_4 is off and T_3 is on, the output is low (logic 0) as shown in Fig. 3.10(b). To ensure that T_4 turns off when T_2 is on diode D_1 is connected in series with T_4.

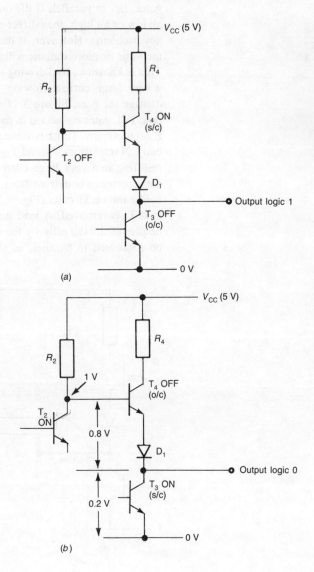

Fig. 3.10

Clamping diodes D_a, D_b and D_c are used to limit any negative voltage excursions that may be present at the input (Fig. 3.9). When the inputs are high, the diodes are reverse biased having no effect on the operation of the circuit.

The totem pole arrangement has low power consumption. This is because the ON transistor conducts through the high impedance of the OFF transistor. Furthermore its output impedance is low which results in a high fan-out capability, in the region of 10.

Open collector (wired-OR) configuration

The totem pole configuration is not suitable for all applications. It is often necessary to connect together the outputs of two or more TTL gates, i.e. in parallel. If the outputs are at the same logic level, i.e. all low or all high, then direct connection is possible without causing any problems. However, if the outputs are at different logic levels, then one or more outputs will be forced to the opposite logic level. This is known as backdriving a gate. If this happens it will result in a very large current flowing in the output transistors which may damage the gate. Figure 3.11 shows the output stages of two totem pole TTL gates connected in parallel with gate A having low and gate B high outputs. If the two outputs are connected to each other, the two ON transistors T_{4b} and T_{3a} will be placed across the supply V_{CC}, resulting in a very large flowing through them which may damage one or both transistors. In order to avoid this, open-collector configuration circuits (Fig. 3.12) are available in which the active load T_4 is removed. A load for the output transistor must therefore be provided externally by the user. The outputs of such gates may be connected in parallel, as shown in Fig. 3.13 in which R_L is a

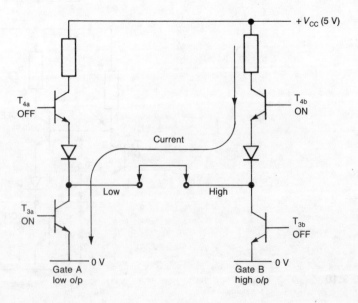

Fig. 3.11

common-load resistor providing the necessary supply voltage to the output transistors of the two gates. This arrangement is known as wired-OR since the output is always low if any of the outputs of gates A, B or C is low.

Low-power TTL

As explained earlier, power dissipation is normally sacrificed in favour of speed by using low-value resistors. For low-power TTL, series 74L, speed is traded off for low-power by using high-value resistors.

Schottky TTL

For fast operation with propagation delay of 9 ns, Schottky transistors may be used. Low-power Schottky, series 74LS, are available with very low power dissipation (2 mW) and relatively good propagation delay of 7 ns. Low-power Schottky normally use diode−transistor logic at the input to reduce power consumption.

Emitter-coupled logic (ECL)

Emitter-coupled logic (ECL) is used when speed is the major consideration. High speed is obtained by preventing the transistors from going into saturation, resulting in a delay time as low as 1 or 2 ns. The operation of an ECL gate is based on the difference amplifier circuit shown in Fig. 3.14. If an output is taken at the collector of T_A, the circuit then behaves as a NOT gate. With a zero input, T_A is turned off with its collector going to V_{CC}, giving a logic 1 output. T_1, however, is forward biased by chain $R_3 - R_4$ ($V_{BB} = V_{CC} \times R_4/[R_3 + R_4]$) whose values are chosen to prevent the transistor from saturating. The T_1 current flows through the common-emitter resistor

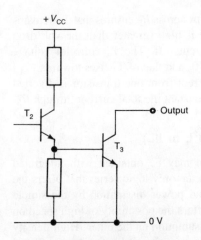

Fig. 3.12 Open-collector gate

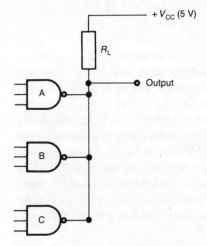

Fig. 3.13 Wired-OR configuration

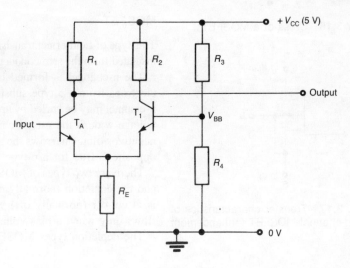

Fig. 3.14 NOT gate using emitter-coupled logic

R_E. The voltage drop that develops across R_E ensures that T_A remains off. When the input to T_A base is high (greater than the volt drop across R_E) T_A conducts and T_1 cuts off. The T_A collector voltage drops, giving an output of logic 0, and that of T_1 rises towards V_{CC}. The process is such that the current from one transistor is diverted to the other transistor keeping constant the total current through R_E.

Integrated injection logic (I²L or IIL)

Very high levels of integration may be obtained with integrated injection logic (I²L or IIL). It is the only logic series that offers the user a choice between speed and power dissipation by the simple choice of external resistors. It offers the possibility of high speed on the one hand and low power consumption on the other. High-density integration is possible with I²L, making it suitable for use in LSI and VLSI.

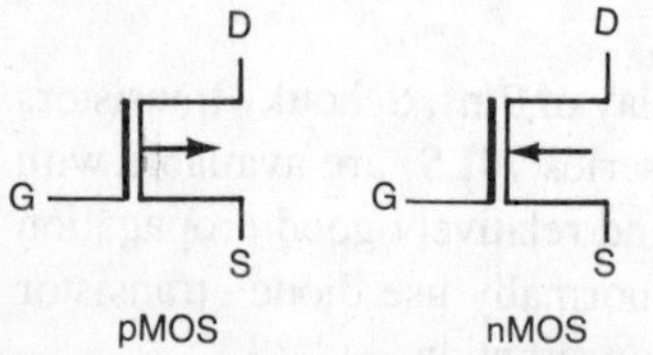

MOS logic

Metal-oxide silicon (MOS) devices are employed in logic gates in a variety of ways. They are based on MOS field-effect transistors (MOSFETs). The field-effect transistor is a unipolar device with one type of carrier, unlike the bipolar transistor which employs two types of carriers. There are two types of MOSFET devices: nMOS and pMOS (Fig. 3.15). The nMOS type has an n-type channel and hence a positive d.c. supply while the pMOS has a p-type channel requiring a negative d.c. supply. Both types may be used for the construction of logic gates. However the nMOS is preferable because it has faster switching than the pMOS type. The most popular arrangement is the complementary MOS (CMOS) which combines both pMOS and nMOS devices.

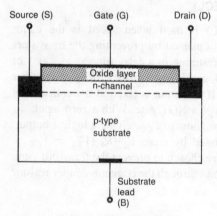

Fig. 3.15 MOSFET symbols

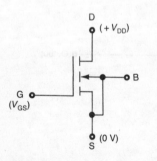

Fig. 3.16 Cross of a MOSFET

MOSFET

This type of field-effect transistor has a metal gate which is electrically insulated from the semiconductor by a thin oxide film. Hence its name.

The n-channel is formed by the gate-insulating oxide attracting electrons from a p-type substrate (Fig. 3.16). The thickness of the n-channel may be varied by applying a voltage to the gate. A positive voltage widens the n-type channel, increasing the current, while a negative voltage narrows the n-channel and reduces the current. The opposite is true for a p-type channel.

There are two types of MOSFETs: the enhancement (normally off) and the depletion (normally on). In the enhancement type the FET is at cut-off (normally off) when the gate bias $V_{GS} = 0$. Current flows only when a bias voltage is applied to the gate (Fig. 3.17).

The depletion types MOSFET conducts (normally on) without a

Fig. 3.17 Transfer characteristics of an n-channel MOSFET (enhancement type)

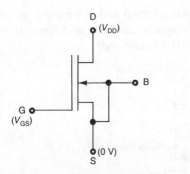

Fig. 3.18 Transfer characteristics of an n-channel MOSFET (depletion type)

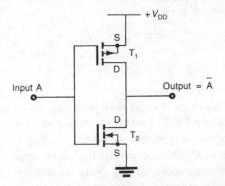

Fig. 3.19 CMOS NOT gate

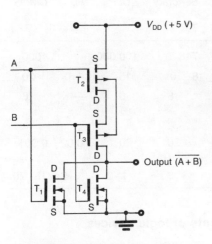

Fig. 3.20 CMOS NOR gate

gate bias. The gate voltage may be positive or negative. For an n-channel type, the drain current increases as gate voltage goes positive and decreases as gate voltage goes negative (Fig. 3.18). A gate-source cut-off voltage V_{GSOFF} which will reduce the drain current to about zero (e.g. -4 V) is normally identified by manufacturers.

Although pMOSFETs are easier to manufacture, the n-type is usually used because of its high density and fast switching.

CMOS

It is possible to reduce power consumption to very small levels (e.g. 50 nW) by using complementary pMOS and nMOS devices. Such devices are of the enhancement type manufactured on the same silicon chip to very high densities. Like all MOS circuits, CMOS devices have very high input impedances and high noise immunity. This family of logic gates (series 4000) finds application in battery-operated portable equipment. Its main disadvantage is its slow speed caused by its high input impedance.

The CMOS NOT gate (Fig. 3.19)

Referring to Fig. 3.19, T_1, being a p-channel MOSFET of the enhancement type, will turn on when its gate, input A, is low, while T_2, being an n-channel MOSFET of the enhancement type, will conduct when its gate, input A, is high. T_1 and T_2 cannot both turn on at the same time. When T_1 is on, T_2 is off and vice versa. The two transistors act as two switches presenting a high impedance 10^{10} Ω when turned off and low resistance (about 1 kΩ) when turned on. When input A is high (logic 1), typically 5 V, T_2 is on, taking the output to earth or zero potential (logic 0). On the other hand, when input A is low, T_1 is on. In this case the output is pulled up to the supply voltage V_{DD} (logic 1).

CMOS NOR gate (Fig. 3.20)

This gate is an extension of the NOT gate in Fig. 3.19 with input A feeding T_2 (p-channel) as well as T_1 (n-channel) and input B feeding T_3 (p-channel) as well as T_4 (n-channel). When either A or B (or both) is high, T_1 or T_4 (or both) will turn on, taking the output down to earth potential (logic 0). Only when A and B are both low will both T_2 and T_3 be on to take the output up to the supply voltage V_{DD} (logic 1). The output therefore has the NOR function F $=$ A$+$B.

The CMOS NAND gate (Fig. 3.21)

When A is low, T_1 is on (and T_4 is off), taking the output to the supply potential V_{DD} (logic 1). Similarly, when B is low, T_2 is on and the output is also high. Only when both A and B are high, will both T_3 and T_4 turn on, taking the output to earth potential (logic 0). The output therefore has the NAND function, AB.

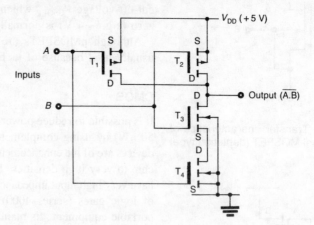

Fig. 3.21 CMOS NAND gate

Other technologies

The main disadvantage of MOS devices is their slow speed of operation. The high-performance MOS, hMOS is a scaled version of nMOS with improved speed and power consumption. High-performance CMOS, known as ChMOS, combines high speed and low power consumption with high packing density. Very fast devices are manufactured under the name of very-fast speed integrated circuits (VFSIC) for military and space applications.

Table 3.1 Comparison of unipolar and bipolar devices

	TTL	TTL low power	Schottky TTL	ECL	I²L	CMOS
Series no.	7400	74L	74S	100 000		4000
Fan-out	8–10	8–10	8–10	15	—	20
Fan-in	8	8	8	5	—	8
Power dissipation (mW)	20	2	20	30	0.02	0.001
Propagation delay (ns)	10	30	3	1–2	1–2	15–20

Power supply requirements of logic devices

Logic devices including TTL and CMOS are designed to operate from a single power supply of a nominal value of +5 V. With TTL devices, a regulated power supply is necessary. This is because the characteristics of TTL devices suffer a substantial change as supply voltage changes. A power supply with a regulation better than 5% must therefore be used to provide a supply voltage falling between 4.75 V and 5.25 V. CMOS devices, on the other hand, are more tolerant of changes in supply voltages and can operate from a supply rail of between 3 V and 15 V. With a current requirement of few μA, CMOS power supplies require very little regulation. Below a supply voltage of 3 V, CMOS continues to operate but with reduced switching speed. TTL devices require considerably more supply current than do their CMOS counterparts. A typical logic gate requires a supply current of about 8 mA, almost 1000 times greater than an equivalent CMOS device. The actual power supply requirements depend on the number of gates used as well as on their type.

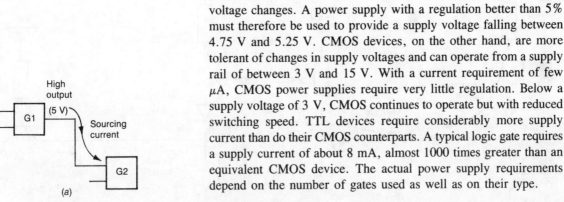

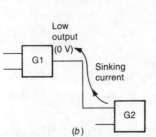

Fig. 3.22 (a) Sourcing current and (b) sinking current

Interfacing of logic devices

When two gates are connected as shown in Fig. 3.22, gate G1 acts as a driver for gate G2. When the output of G1 is high, current flows into G2 as shown in Fig. 3.22(a). This is known as a sourcing current, which must be large enough to drive G2. When the output of G1 is low, current flows into the driver from G2 as shown in Fig. 3.22(b). This current is known as a sinking current. In this case, the driving gate must be able to handle the necessary sinking current without changing its logic state.

TTL to TTL

Consider the TTL circuit in Fig. 3.23 showing the output stage of a driving gate connected to one of the inputs of a similar gate. In Fig. 3.23(a) the output of the driver is high with T_1 on and T_2 off. T_1 provides the sourcing current I_{sc} to drive T_3. But T_3, being off, requires just the leakage current I_l rated at 40 μA for a standard TTL device. Given that the maximum rating of the sourcing current I_{sc} that T_1 can provide is 400 μA, it follows that the maximum number of TTL units that may be driven by the TTL gate when its output is high is

$$\text{fan-out high} = \frac{\text{Maximum sourcing current } I_{sc}}{\text{leakage current } I_l} = \frac{400 \ \mu A}{40 \ \mu A}$$

$$= 10.$$

In Fig. 3.23(b) the output of the driver is low; T_1 is off and T_2 is on. The sinking current I_{sk} flows into T_2 as shown. T_3 is now on with a saturating current I_{sat} requirement of 1.6 mA. However, the maximum sinking current that may be provided by T_2 is 16 mA, giving

MATTHEW BOULTON
COLLEGE LIBRARY

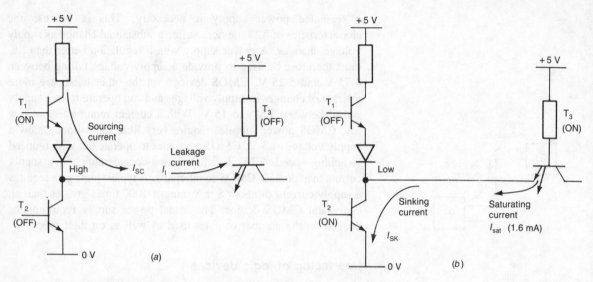

Fig. 3.23 TTL driving TTL

$$\text{fan-out low} = \frac{\text{Maximum sinking current } I_{sk}}{\text{Saturating current } I_{sat}} = \frac{16 \text{ mA}}{1.6 \text{ mA}}$$
$$= 10.$$

In this case both fan-out high and fan-out low are the same. However, this may not be the case for all devices, in which case the lower figure for fan-out must be used to avoid overloading the driver.

If overloading occurs then either the output-high voltage drops below its threshold level of 2 V or the output-low voltage rises above its threshold level of 0.8 V. These threshold levels assume no noise in the signal. If noise is taken into account, the figures for fan-out will be different.

CMOS to TTL

In some applications, it is desirable to use two types of logic, combining the properties of both; TTL for speed and CMOS for low power consumption. Figure 3.24 shows a CMOS gate driving a TTL device. Since CMOS can function with a supply voltage of $3-15$ V, it can function adequately with a regulated 5 V supply necessary for TTL devices. When CMOS output is high (Fig. 3.24(a)), T_1 is on with either or both T_2 and T_3 off. Sourcing current I_{sc} is supplied by T_1 to provide the small leakage current I_1 required by T_4. The maximum value of I_1 is 40 μA, presenting no problem to the driving CMOS gate. When the output of the CMOS gate is low (Fig. 3.24(b)), T_1 is off and both T_2 and T_3 are on. A sinking current I_{sk} now flows into T_2 and T_3 supplied by the ON TTL input transistor T_4. This means that the CMOS device has to sink a current of 1.6 mA through T_2 and T_3 as shown. A voltage will thus develop across the CMOS output determined by the resistance of the two ON transistors, T_2 and T_3. This voltage could be high enough for the output to cross the

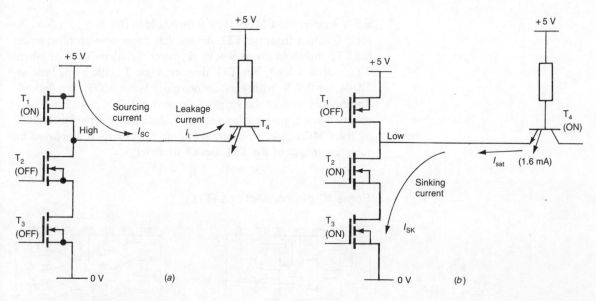

Fig. 3.24 CMOS driving TTL

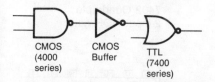

CMOS
(4000
series)

CMOS
Buffer

TTL
(7400
series)

Fig. 3.25 The use of a CMOS buffer

low-threshold level of 0.8 V (or 0.4 V if noise margin is taken into consideration). For this reason, devices known as CMOS buffers are sometimes employed between CMOS and TTL gates, as shown in Fig. 3.25. Compatible CMOS devices are also available with a built-in buffer. These ICs are given the prefix B, e.g. the 4011B quad two-input buffered NAND gate.

TTL to CMOS

Now consider a TTL gate driving a CMOS device in Fig. 3.26. The low-level threshold for CMOS is 30% of the supply voltage. Using

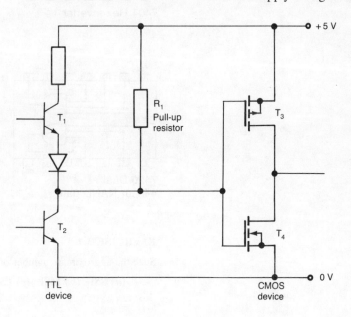

Fig. 3.26 TTL driving CMOS

a 5 V supply, the CMOS logic 0 threshold is $0.3 \times 5 = 1.5$ V. A logic 0 output from the TTL device thus creates no problem since the TTL threshold low is 0.8 V. However, problems do arise when TTL output is high. For TTL devices, logic 1 could be as little as 2.0 V (or 2.4 V with a noise margin). For CMOS, the logic 1 threshold is 70% of the supply voltage, i.e. $0.7 \times 5 = 3.5$ V. To ensure that a logic high from the TTL device is accepted as logic 1 by the CMOS gate, an external pull-up resistor R_1 is employed to raise the output of the TTL device to about 5 V.

Some IC pin connections (TTL)

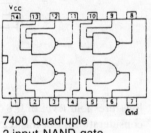

7400 Quadruple
2-input NAND gate

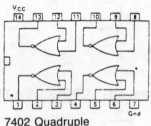

7402 Quadruple
2-input NOR gate

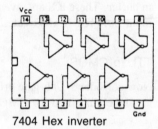

7404 Hex inverter

7408 Quadruple
2-input AND gate

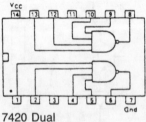

7420 Dual
4-input NAND gate

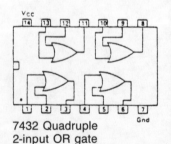

7432 Quadruple
2-input OR gate

EXERCISE 3.1

State the approximate number of gates contained in each of the following:

(*a*) SSI (*b*) MSI (*c*) LSI (*d*) VLSI.

EXERCISE 3.2

List typical values for each of the following terms used in relation to standard TTL logic gates:

(a) Logic 0 threshold (b) Logic 1 threshold (c) Fan out (d) Fan in (e) Supply voltage (f) Propagation delay.

EXERCISE 3.3

Name a logic family suitable for each of the following:

(a) large supply voltage range and low supply current
(b) propagation delay less than 2.5 ns
(c) slow, very high packing density, very low supply voltage and low power dissipation
(d) logic 1 level of approximately 3.5 V and logic 0 level of approximately 0.3 V with a supply of 5 V.

EXERCISE 3.4

A logic circuit is to be used in a portable electronic equipment powered by a small 9 V supply battery. The circuit is to be driven with a frequency of 100 kHz. Name a suitable logic family for this application and give its series number.

EXERCISE 3.5

Give the meaning of the following terms:

fan-in, fan-out, threshold voltage, open collector, noise margin, sourcing current, sinking current.

EXERCISE 3.6

Referring to the 7410 IC

(a) name the IC stating its logic family
(b) state the recommended power supply and a typical propagation delay
(c) state the number of similar gate inputs that may be driven by one gate 7410 output.

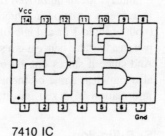

7410 IC

EXERCISE 3.7

Explain the difference between the following TTL families:

74, 74H, 74L, 74S, 74LS

4 Bistables and shift registers

So far we have only considered combinational logic systems in which the output is determined by the combination of the inputs present at the time. Such systems have no memory. Sequential systems on the other hand have outputs that are dependent upon the order of the input signals as well as on the actual input present at that particular time. Such systems have a memory capability and are synchronised by a clock pulse.

The one-bit memory or latch

A basic latch circuit is shown in Fig. 4.1. It consists of two NOT gates (single-input NAND gates) G1 and G2 with the output of each gate fed back to the input of the other. This feedback combination is called a flip-flop. The most important property of a flip-flop circuit is that it has only two stable states ($Q = 1$, $\overline{Q} = 0$ and $Q = 0$, $\overline{Q} = 1$). For example if the output of G1 is $Q = 1$, then B, the input to G2 is at logic 1 as well. Gate G2 being an inverter will have its output $\overline{Q}$ at logic 0. Since $\overline{Q}$ is tied to A, then the input to G1 is also 0 and its corresponding output, Q is at logic 1, which conforms with our starting point. Hence $Q = 1$, $\overline{Q} = 0$ is one possible steady state. Similarly, it can be proved that $Q = 0$, $\overline{Q}$ is another possible state. However a situation in which outputs Q and $\overline{Q}$ are in the same state (both at logic 1 or at logic 0) is not possible. The flip-flop has only two steady states hence it is also known as a binary or a bistable circuit. And since it stores one bit of information ($Q = 1$ or $Q = 0$) it is also known as a one-bit memory unit or cell. Furthermore, since the information is locked into it, it acts as a latch.

S–R flip-flop

The basic circuit in Fig. 4.1 does not provide for setting the output to any particular state. To do this two more NAND gates G3, G4 are added to make the set–rest (S–R) flip-flop (or bistable) as shown in Fig. 4.2. If it is now desired to store $Q = 1$, then the circuit is

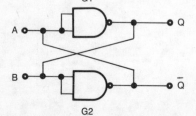

Fig. 4.1 Basic flip-flop or latch

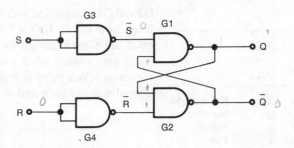

Fig. 4.2 S–R flip-flop using NAND gates

Table 4.1

S	R	Q_{n+1}	
0	0	Q_n	No change
0	1	0	Reset
1	0	1	Set
1	1	Indeterminate	

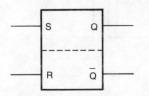

Fig. 4.3 S–R flip-flop symbol

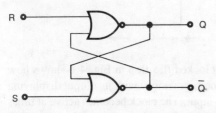

Fig. 4.4 S–R flip-flop using NOR gates

set by making input S = 1 and input R = 0. With S high, the output of G3 is 0 making Q = 1 (when one input of a NAND gate is low, its output is high) and that in turn forces one of the G2 inputs to logic 1 as well. With R = 0, G4 output is at 1. Gate G2 now has two inputs both at logic 1 giving an output $\overline{Q}$ = 0. Conversely to store Q = 0, the circuit is reset with R = 1 and S = 0 giving an output of Q = 0 and $\overline{Q}$ = 1.

The full truth table for an S–R flip-flop is shown in Table 4.1. As stated earlier, the output of sequential systems depends on the previous state of the inputs as well as on its present state. For this reason Q_n is used to denote the previous state of Q, while Q_{n+1} denotes its current state. A combination of S = R = 0 leaves the output unchanged (Q_n). Combination S = R = 1 is indeterminate and hence not allowed. Gates G1 and G2 by themselves form what is known as $\overline{S}$–$\overline{R}$ flip-flop in which case $\overline{S}$ = 0, $\overline{R}$ = 1 will set and $\overline{S}$ = 1, $\overline{R}$ = 0 will reset the flip-flop. Figure 4.3 shows the logic symbol for the simple S–R bistable.

A flip-flop circuit can also be constructed using two NOR gates as shown in Fig. 4.4.

The clocked S–R flip-flop

In sequential systems it is often required to set or reset a flip-flop in synchronism with other units and in accordance to a clock. To achieve this, the clocked or triggered S–R circuit shown in Fig. 4.5 is utilised. When the clock is at logic 0, i.e. between clock pulses,

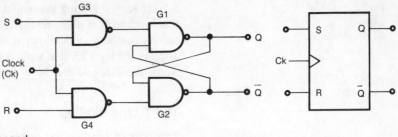

Fig. 4.5 (a) Clocked S–R flip-flop, and (b) clocked S–R symbol

(a)

(b)

Table 4.2

Clock (Ck)	(t_n)		(t_{n+1})	
	S	R	Q_{n+1}	
0	x	x	Q_n	No change
1	0	0	Q_n	No change
1	0	1	0	Reset
1	1	0	1	Set
1	1	1		Indeterminate

G3 and G4 are disabled with their outputs high regardless of the logic levels of S and R. If Q is 1 it remains at logic 1 and vice versa. In other words the bistable does not change states between clock pulses; it can only change when a clock pulse is present. When the clock is present (clock pulse at logic 1) the S−R flip-flop functions in the normal way, as indicated in Table 4.2.

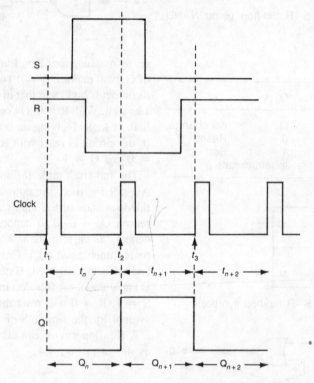

Fig. 4.6   Timing diagram for a clocked flip-flop

The timing diagram for a clocked flip-flop in Fig. 4.6 shows how the S−R flip-flop acts as a one-bit memory with its output displaying the last signal received at the input. The clock becomes active at times t_1, t_2, t_3, etc., with t_n representing the time interval before the clock pulse and t_{n+1} representing the time period after the clock pulse. At active clock pulse t_1, S = 0 and R = 1, giving an output $Q_n = 0$ over the time interval t_n. At t_2, the S−R combination is changed to S = 1, R = 0 forcing the output Q_{n+1} to change to logic 1. At t_3, S is low and R is high, resetting the flip-flop to $Q_{n+2} = 0$, and so on. As can be seen, the output at any time interval (Q_{n+1} at t_{n+1}) displays the logic level at input S prior to the time interval. In other words the data at S has been delayed for one time interval.

The T-type flip-flop

This type of flip-flop has one T (toggle) input in which the output changes states in response to every input pulse. A T-type flip-flop

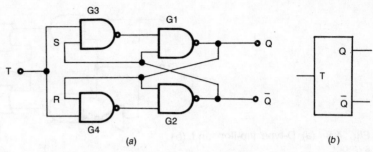

Fig. 4.7 (a) T-type flip-flop, and (b) symbol

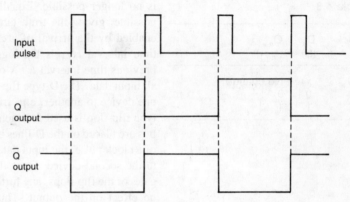

Fig. 4.8 Timing diagram for a T-type flip-flop

may be constructed by applying feedback from Q to R and from $\overline{Q}$ to S as shown in Fig. 4.7. Assuming the flip-flop is initially set with Q = 1 and $\overline{Q}$ = 0, then S = 0 and R = 1, because of the feedback. With input T at 0, G3 has both its inputs at 0, resulting in a logic 1 at its output. G4 on the other hand has one input at logic 0 and the other at logic 1, giving a logic 1 at its output. When the input is pulsed to logic 1 it changes one of the inputs to G4 from 1 to 0, resulting in a change of the state of its output from 0 to 1. G2 now has one of its inputs at logic 0 forcing the output $\overline{Q}$ to change to 1. G1 on the other hand has both its inputs high, giving an output Q of 0. A similar process will take place when the next pulse arrives at the input. The output alternates (toggles) between logic 1 and logic 0 in response to each input pulse. The output therefore changes at half the frequency of the input, as shown in Fig. 4.8.

D-type flip-flop

The clocked S−R may be modified as shown in Fig. 4.9 to provide a one-bit delay for a single input data line, D. An inverter is added to the R input so that the R input is the complement (inverse) of the S input. In this way the flip-flop is always in the set state (D = 1) or the reset state (D = 0). Basically the circuit is an S−R flip-flop with the ambiguous state of S = R = 1 removed since this condition

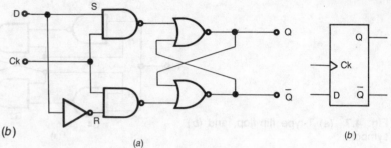

Fig. 4.9 (a) D-type flip-flop and (b) symbol

Table 4.3

Clock	D_n (t_n)	Q_{n+1} (t_{n+1})
0	x	No change
1	0	0
1	1	1

is no longer possible. Equally the state $S = R = 0$ is no longer possible, giving the truth table shown in Table 4.3. The circuit is enabled by the arrival of a clock pulse. The output (Q_{n+1}) during time interval t_{n+1} is the logic state of the input (D_n) during the previous time interval t_n. A one-bit time delay is thus achieved for all input data. The D-type flip-flop is useful in transferring data from one device to another, e.g. from a memory to a register where a D-type flip-flop is used for each bit as shown in Fig. 4.10. The input bits are placed on the D-lines by the input device, and with the arrival of a clock pulse, the input data are transferred to outputs Q_0-Q_3 and to the second device. Once the clock pulse has triggered the input gates of the flip-flops, any further changes in the input states will have no effect on the output. Thus the time interval t_n between clock pulses is available for placing a batch of data on the D-lines.

The J–K flip-flop

The J–K flip-flop is similar to a T-type with two additional inputs, as shown in Fig. 4.11. These additional input are known as J and K to distinguish them from S and R. The J–K construction provides a universal programmable flip-flop.

With $J = 1$ and $K = 0$, the flip-flop is in the set condition ($Q = 1$, $\overline{Q} = 0$). The arrival of a clock pulse forces the clock input to G3 and G4 to go to logic 1 which causes G3 output to change from 0 to 1 and with it the output of G1 (Q) to change from 1 to 0. This will, in turn change the output of G2 ($\overline{Q}$) from 0 to 1. If now the input combination changes to $J = 0$ and $K = 1$, then the arrival of a clock pulse will change the output of G4 from 1 to 0 and with it outputs $\overline{Q}$ from 1 to 0 and Q from 0 to 1.

When $J = K = 0$, each of NAND gates G3 and G4 has at least one input at logic 0. Their outputs are therefore held at logic 1, keeping the logic states of Q and $\overline{Q}$ unchanged regardless of the arrival of a clock pulse.

The condition of $J = K = 1$, which is forbidden in the S–R flip-flop, is permitted with the J–K type. Its effect is to make the bistable behave like a T-type flip-flop. While the clock pulse is at logic 0, G3 and G4 are disabled and no change in the outputs can take place.

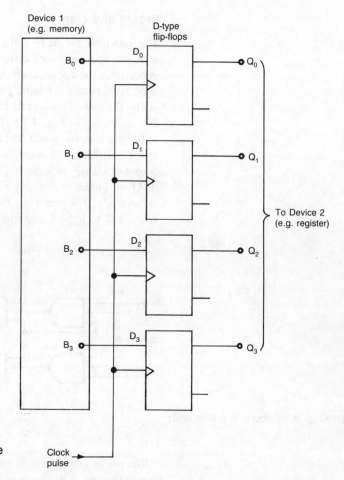

Fig. 4.10 Data transfer using D-type flip-flops

Table 4.4

Clock (Ck)	J	K	Q_{n+1}	
0	x	x	Q_n	No change
1	0	0	Q_n	No change
1	1	0	1	
1	0	1	0	
1	1	1	Toggle	

When a clock pulse is present (logic 1), G3 and G4 are enabled with their outputs determined by the feedback inputs from Q and $\overline{Q}$. Since Q is fed to G4 and its complement $\overline{Q}$ is fed back to G3, it follows therefore that output Q (and $\overline{Q}$) will alternate or toggle between 1 and 0 with the arrival of each clock pulse.

Table 4.4 shows the complete truth table for a J–K flip-flop.

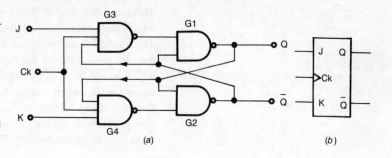

Fig. 4.11 (a) J–K flip-flop and (b) symbol

Preset and clear

The truth table for the J−K flip flop describes how the output changes with the application of different combinations of the input. It is important in most applications to be able to set the output to a particular logic level before the input pulses are applied. This may be carried out by the introduction of two additional inputs to latch gates G1 and G2, as shown in Fig. 4.12. The flip-flop is **preset** ($Q = 1$ and $\bar{Q} = 0$) by setting the preset (Pr) input to logic 0 and the clear (Cr) input to logic 1. This will drive the output of G2 ($\bar{Q}$) to logic 0 and the output of G1 (Q) to logic 1. Conversely, to **clear** the flip-flop ($Q = 0$, $\bar{Q} = 1$) the preset and clear inputs are set to logic 1 and 0, respectively. The preset and the clear are active low, i.e. a low preset will set Q to logic 1 and vice versa.

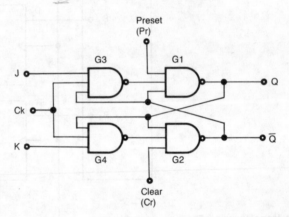

Fig. 4.12   J−K flip-flop with preset and clear

The preset and clear data are called direct or asynchronous inputs due to the fact that they may be applied at any time between clock pulses and as such are not in synchronism with the clock. However once the state of the flip-flop is established, both of the direct inputs (preset and clear) must be maintained at logic 1 to allow the flip-flop to respond to the next pulse. This condition is known as the enable condition which allows the flip-flop to function in the synchronous mode.

Race-around condition

There is a practical problem with the J−K flip-flop described above which may cause race-around or oscillations at every clock pulse. This occurs if the width of the clock pulse is large compared with the switching time of the flip-flop. Under this condition the output which is fed back to the input changes the latter, causing a further change in the output and so on until the end of the clock pulse, making the final output of the flip-flop ambiguous. One way of overcoming this problem is to use very narrow clock pulses. However with switching times for modern ICs improving rapidly, such a technique may not be adequate. There are two other techniques that may be used to avoid the problem of unstable operation: edge triggering and master—slave flip-flops.

Edge triggering

A pulse has three component parts, as shown in Fig. 4.13, namely a leading edge, an active level and a trailing edge. For a positive-going pulse the leading edge is positive and the trailing edge is negative. The opposite is true for a negative-going pulse. There are two main ways of using a pulse such as a clock pulse for triggering purposes:

1. level triggering in which a bistable may change state in response to its input throughout the duration while the clock pulse is held high for positive-going pulses or low for negative-going pulses; and

2. edge triggering in which a bistable changes state in response to a transition of the clock pulse from 0 to 1 (positive edge triggering) or from 1 to 0 (negative edge triggering). The flip-flop is thus enabled only for the duration of the transition.

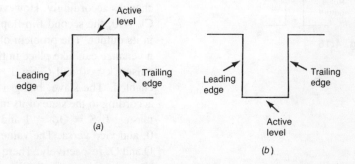

Fig. 4.13

Edge triggering may be achieved by the use of the simple *RC* differentiating circuit shown in Fig. 4.14. Another technique employing two gates is shown in Fig. 4.15, in which the propagation delay of the NAND gate is used to produce a very narrow pulse. When the clock input changes from logic 0 to 1, both inputs to G2 are held

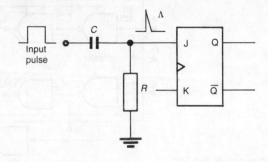

Fig. 4.14

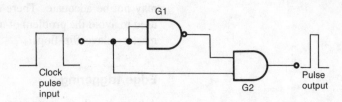

Fig. 4.15

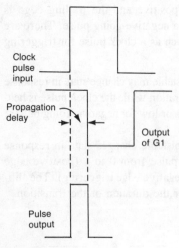

Fig. 4.16

at logic 1 for a short duration equal to the time it takes the clock pulse to propagate through gate G1, producing a pulse duration equal to the propagation delay of G1 (Fig. 4.16).

Master–slave flip-flop

By far a better method of avoiding the race-around condition is the use of the master–slave flip-flop shown in Fig. 4.17. A cascade of two S–R bistables is connected as shown with the output of the second bistable (known as the slave) being fed back to the input of the first (known as the master). Positive clock pulses are applied to the master and these are inverted by G1 before being fed to the slave. When a clock pulse is applied (Ck = 1) the master is enabled and its output changes accordingly. However, the clock pulse going to the slave is $\overline{Ck} = 0$, the second flip-flop is disabled and no change takes place in its output. The problem of oscillation is thus circumvented since no change can take place in the output state. At the end of the clock pulse Ck = 0 and $\overline{Ck} = 1$, the master is disabled and the slave is enabled. The slave, being a S–R flip-flop, will change its output according to the state of its inputs which have been set earlier by the master. If $S = Q_M = 1$ and $R = \overline{Q}_M = 0$, then Q = 1 and $\overline{Q} = 0$, and vice versa. The value of Q_M and $\overline{Q}_M$ are thus transferred to Q and $\overline{Q}$, respectively. There is no question of oscillation in this case since there is no feedback to the S–R inputs of the slave. In short, during a clock pulse the output Q does not change but Q_M follows the J–K logic; at the end of the clock pulse, the value of Q_M is transferred to Q.

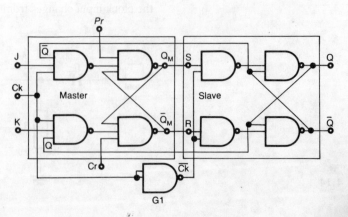

Fig. 4.17 Master–slave J–K flip-flop

Table 4.5

	Number	Contents	Max. frequency (MHz)
TTL	7473	Dual J–K	20
	7474	Dual D-type	25
CMOS	4013	Dual D-type	8
	4027	Dual J–K	5

TTL and CMOS flip-flops

Flip-flops are manufactured using TTL (74 series) and CMOS (4000 series) technologies. Table 4.5 shows some of the most popular IC packages.

Shift registers

It is often necessary in digital electronics to provide temporary storage for data before transition for further processing. The basic storage element is the flip-flop. Each flip-flop stores one bit of data. Therefore to store an n-bit data word, n flip-flops are necessary, constructed so as to form a shift register. Binary data may be transferred in one of two modes: the serial and parallel modes (Fig. 4.18). In the serial mode the bits are transferred in sequence, one after the other: b_0, b_1, b_2, and so on. In the parallel mode the bits are transferred simultaneously along a number of parallel lines (four lines for a four-bit word) in synchronism with a single pulse from the system clock. There are thus four basic ways in which a shift register may be used to store and transfer data from one part of a system to another:

1. serial input to parallel output (SIPO);
2. serial input to serial output (SISO);
3. parallel input to serial output (PISO); and
4. parallel input to parallel output (PIPO).

Serial in — parallel out (SIPO)

A four-bit shift register is shown in Fig. 4.19 employing four flip-flops of the S–R (or J–K) master–slave type. Note that the input flip-flop FF3 is converted to a D-type flip-flop, allowing a series of binary input bits to be fed into the register. Assuming the four-bit serial input 1011 with the most significant bit (left-most digit) $b_3 = 1$ and the least significant bit (the right-most digit) $b_0 = 1$. Before the input is fed into the shift register the flip-flops are cleared by applying a logic 0 to the clear Cr input so that every output $Q_0 - Q_3$ is set 0. Cr is then set to logic 1 to enable the flip-flops. The serial data train and the synchronous clock are now applied. After the first clock pulse, the least significant bit b_0 is entered into FF3, changing

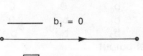

(a)

$b_0 = 1$

$b_1 = 0$

$b_2 = 1$

$b_3 = 1$

(b)

Fig. 4.18 Data transfer: (a) serial mode, and (b) parallel mode

MATTHEW BOULTON
COLLEGE LIBRARY

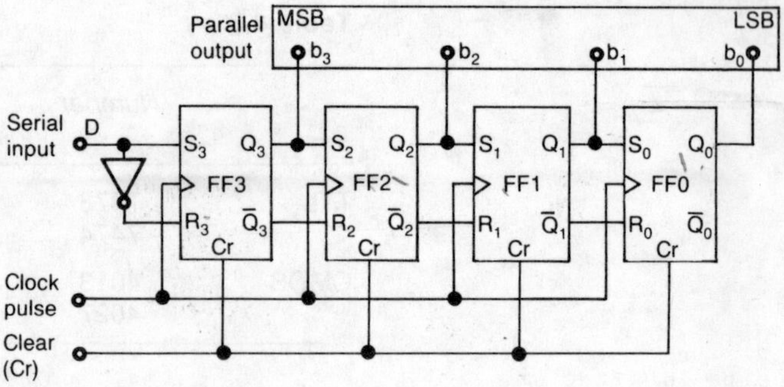

Fig. 4.19 Serial in–parallel out

Table 4.6

Clock pulse	*Input* Bit	*Logic*	*Output after each clock pulse* Q_3	Q_2	Q_1	Q_0
0	*x*	*x*	0	0	0	0
1	b_0	1	b_0	0	0	0
2	b_1	0	b_1	b_0	0	0
3	b_2	1	b_2	b_1	b_0	0
4	b_3	1	b_3	b_2	b_1	b_0
			Parallel output			

Q_3 from logic 0 to logic 1 while all other outputs remain at 0. After the second clock pulse, logic 1 at S_2 is transferred to Q_2, and $b_1 = 0$ of the input data is entered into FF3. No other change takes place in the other bistables. After clock pulse 3, b_2 is entered into FF3 with the other bits at S_2 and S_1 shifted to Q_2 and Q_1, respectively (as shown in Table 4.6), resulting in $Q_1 = 1$, $Q_2 = 1$ and $Q_3 = 0$. Q_0 remains at logic 0. After clock pulse 4, the MSB is entered into FF3, shifting the other bits across and giving the following output: $Q_3 = 1$, $Q_2 = 0$, $Q_1 = 1$, $Q_0 = 1$. The input word is thus installed in the register with each output being available on a separate line (b_0, b_1, b_2, b_3). A parallel output has thus been produced.

Serial in — serial out

Once the input data are stored in their appropriate flip-flops in the manner described above, they may be reproduced in a serial form. This is done by taking the output at Q_0 which is at logic $b_0 = 1$. A clock pulse is then applied thus shifting the bits one place to the right with Q_0 acquiring logic level $b_1 = 0$. Another clock pulse is then applied, and so on. After four clock pulses the input reappears at the output in a serial form.

Parallel in — serial out

Figure 4.20 shows a four-bit shift register that may be used as a parallel-to-serial converter. The parallel input is fed into preset inputs

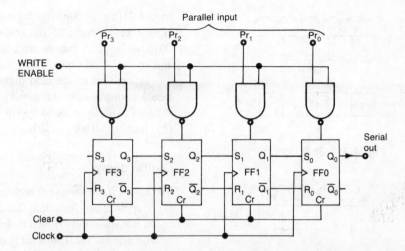

Fig. 4.20 Parallel in–serial out

Pr_0, Pr_1, ..., Pr_3 to set each flip-flop to the appropriate logic level. Consider the parallel four-bit input b_0-b_3. The LSB b_0 is applied to Pr_0; b_1 is applied to Pr_1, and so on, so that $Pr_0 = 1$, $Pr_1 = 1$, $Pr_2 = 0$ and $Pr_3 = 1$. A logic 1 is then applied to the clear (Cr) input line. This will clear the flip-flops. Once the register is cleared, the Cr line is taken to logic 1 and maintained at that level to enable the flip-flops.

Similarly, a logic 1 at the WRITE ENABLE allows the parallel input to be fed or written into the flip-flops. The output of each flip-flop will thus be set to the logic level present at the input. The parallel input bits are therefore transferred and are stored into the appropriate flip-flop.

To obtain the stored word in a serial form, clock pulses are applied to shift the stored bits to the left. After four clock pulses, the input word reappears at the output Q_0 in a serial form.

Parallel in — parallel out

Once the input data bits are stored into their appropriate flip-flops, they are then available in parallel form at outputs Q_3, Q_2, Q_1 and Q_0.

Shift right and shift left

It is often necessary to shift the contents of a register one or more bits to the left or to the right to perform data manipulation or mathematical operations.

Consider the contents of the shift register in Fig. 4.19 with the least significant bit stored in FF0. Now, if one clock pulse is applied, then a shift to the right occurs with each bit moving to the next lower significant place, thus performing a binary division by 2. Provided the serial input is held low, then FF3, which holds the MSB will be low when the shift right operation is carried out. Consider, for example, the contents of a five-bit register, $01010_2 = 10_{10}$ shown in

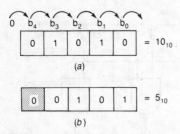

Fig. 4.21 Shift right (binary division)

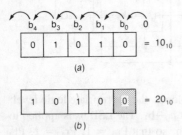

Fig. 4.22 Shift left (binary multiplication)

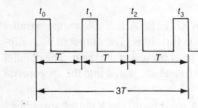

Fig. 4.23

Fig. 4.21(*a*). By shifting the contents one place to the right, **shift right**, the contents of the register change to that shown in (*b*), $00101_2 = 5_{10}$. A further shift to the right divides the number by 4 and so on. A one-bit shift to the left, **shift left**, on the other hand moves each bit to the next higher significant digit (Fig. 4.22) and hence multiplies the number by 2. For the number shown in Fig. 4.22(*a*) ($01010_2 = 10_{10}$) a shift left produces the binary number in (*b*), namely $10100_2 = 20_{10}$.

Digital delay

The basic circuit for a shift register (Fig. 4.19) may be used for several other applications including that of a digital delay line. If one-bit data is fed into the input D, then at the first active edge of the clock pulse at time t_0 (Fig. 4.23), this bit is entered into flip-flop FF3. After one complete clock cycle, at time t_1, the data bit is shifted to FF2, and so on. At time t_3 the one-bit data input appears at output Q_0. As can be seen a time delay of $3T$ is introduced, where T is the period of the clock pulse. A similar delay will be introduced for a multi-bit input. In general, a pulse train suffers a delay of $(n-1)T$ when fed through an n-stage shift register.

Some TTL IC pin connections

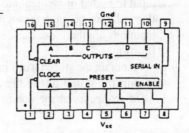

7496 5-bit
shift register

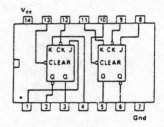

74107 Dual JK flip-flop

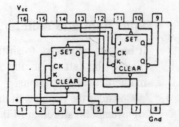

74109 Dual JK positive
edge-triggered flip-flop

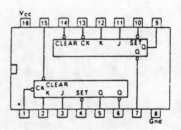

74112 Dual JK triggered
flip-flop

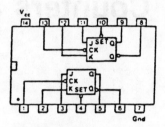

74113 Dual JK negative-
triggered flip-flop

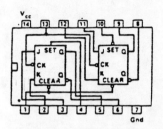

74114 Dual JK negative
edge-triggered flip-flop

EXERCISE 4.1

(*a*) Draw a circuit diagram to show how two NAND gates may be used to form an R−S bistable and label the inputs.

(*b*) Construct a truth table for the circuit in (*a*). Give one application for an R−S bistable.

EXERCISE 4.2

Draw a logic diagram of a 4-bit shift register with the following facilities: serial input; serial output; parallel input; parallel output.

EXERCISE 4.3

(*a*) Explain how binary multiplication may be obtained by a 'shift left' technique.

(*b*) Draw a circuit diagram which may be used to multiply a 4-bit number by 2.

EXERCISE 4.4

(*a*) Construct a truth table for a J−K flip-flop.

(*b*) State what is meant by 'master slave' and the reason for its use.

(*c*) Show how a J−K flip-flop may be converted to a D-type flip-flop.

EXERCISE 4.5

(*a*) A shift register is loaded with 11110000 and its data input is at logic 0. If the data is shifted from left to right, state the binary code contained in the register after TWO shift pulses.

(*b*) Sketch a logic circuit to show how THREE JK bistables can be connected together to form a 3-stage, serial-in/serial-out, shift register with a reset facility.

EXERCISE 4.6

In a 4-bit SISO shift register, state the number of clock pulses required to shift one byte of data from input to output.

5 Counters and timers

Bistables or flip-flops may be used as binary dividers (or divide-by-2, ÷ 2 devices) to form a counter. A counter is a group of flip-flops arranged so that they indicate the total number of pulses applied to the input. Counters may be divided into two broad categories: asynchronous and synchronous.

Asynchronous counters

Consider a chain of four J–K master–slave flip-flops with output Q of each stage connected to the clock input of the following stage as shown in Fig. 5.1. With J and K of each stage tied to the d.c. supply (logic 1), each flip-flop functions as a T-type bistable. Each stage will thus change the state of its output at every active edge of the clock pulse, as shown in Fig. 5.2. The flip-flops are first reset making $Q_0 = Q_1 = Q_2 = Q_3 = 0$. Pulses to be counted are then applied to the clock input of FF0 and assuming a negative active edge, then at time t_1, the first active clock edge Q_0 changes states from logic 0 to logic 1. This change has no effect on the following flip-flops, keeping their Q outputs at logic 0 (also see Table 5.1). At t_2, Q_0 changes back to 0. This negative transition is fed to the clock input of the following flip-flop FF1, changing the state of its output to logic 1. FF2 and FF3 remain unaffected. At time t_3, Q_0 changes to logic 1, and that being a positive transition, it has no effect on the following flip-flop. At time t_4, Q_0 again changes state, from logic 1 to logic 0. This causes FF1 output Q_1 to change from 1 to 0, and this being a

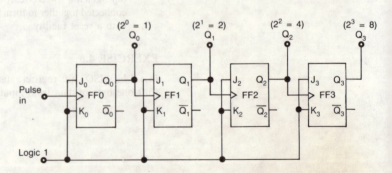

Fig. 5.1 Ripple-through (asynchronous) counter

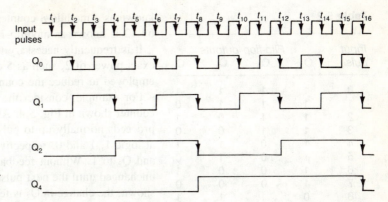

Fig. 5.2 Timing diagram for a ripple-through counter

Table 5.1

Input pulse	Flip-flop outputs			
	Q_3	Q_2	Q_1	Q_0
0	0	0	0	0
1	0	0	0	1
2	0	0	1	0
3	0	0	1	1
4	0	1	0	0
5	0	1	0	1
6	0	1	1	0
7	0	1	1	1
8	1	0	0	0
9	1	0	0	1
10	1	0	1	0
11	1	0	1	1
12	1	1	0	0
13	1	1	0	1
14	1	1	1	0
15	1	1	1	1
16	0	0	0	0

negative transition causes FF2 output Q_2 to change from logic 0 to logic 1, and so on at each active edge of the input pulses. Negative transitions 'ripple' through the counter from FF0 (LSB) to FF3 (MSB), hence the name, 'ripple-through' counter.

As can be seen from Table 5.1, outputs Q_0, Q_1, Q_2, and Q_3 give the binary representation of the number of pulses fed into the input of the counter. Up to $2^4 = 16$ pulses (pulse 0 to pulse 15) may be counted using four flip-flop stages. The counter resets itself at the sixteenth pulse, as shown. In general, a chain of flip-flops counts in the binary system up to the number 2^{N-1} before it resets itself to its original state where N is the number of flip-flop stages in the counter.

Up and down counters

Referring back to the counter in Fig. 5.1 and to Table 5.1, it can be seen that with each input pulse, the count is increased or incremented by 1. The count is in ascending order, producing what is known as an **up-count**. This was the direct result of selecting the Q output of each flip-flop. A reverse count or a **down-count** may be produced in which the count is decreased or decremented by 1 with each input pulse (Table 5.2). This is achieved by selecting the $\overline{Q}$ outputs of the flip-flops.

A counter may be made to count in either the forward or reverse direction. Such counters are known as programmable or up/down counters.

Divide-by-*n*-counter

The basic binary counter shown in block diagram form in Fig. 5.3 counts up to $2^3 = 8$ pulses (0 to 7). If a single output is taken at C, then the counter becomes a divide by $2^3 = 8$. In other words, one output pulse will be produced at C for every eight pulses at the input. A four-stage counter will divide by $2^4 = 16$, and so on. In general, a counter with N stages will divide by $n (= 2^N)$. Such a counter is

Table 5.2

Input pulse	Flip-flop outputs			
	$\bar{Q}_3$	$\bar{Q}_2$	Q_1	$\bar{Q}_0$
0	1	1	1	1
1	1	1	1	0
2	1	1	0	1
3	1	1	0	0
4	1	0	1	1
5	1	0	1	0
6	1	0	0	1
7	1	0	0	0
8	0	1	1	1
9	0	1	1	0
10	0	1	0	1
11	0	1	0	0
12	0	0	1	1
13	0	0	1	0
14	0	0	0	1
15	0	0	0	0
16	1	1	1	1

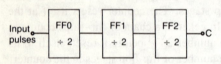

Fig. 5.3   Divide-by-8 counter

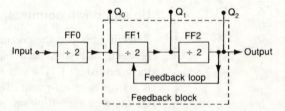

Fig. 5.4   Counter employing feedback

Table 5.3

Input pulse	Q_2	Q_1	Q_0
0	0	0	0
1	0	0	1
2	0	1	0
3	0	1	1
4	1	0	0
Feedback	1	1	0
5	1	1	1
6	0	0	0

known as a modulo-*n* counter. For example, a counter that divides by 12 is known as modulo-12 counter, and so on.

It is frequently necessary to divide by a number which is not an exact power of 2, such as 5 or 10. In such cases feedback must be employed to reduce the count.

For example, consider the feedback applied to the three-element counter shown in Fig. 5.4. As can be seen from Table 5.3 the count proceeds normally up to pulse number 3 when Q_0, Q_1 and Q_2 are at logic 1, 1 and 0, respectively. Pulse 4 changes Q_0 to 0, Q_1 to 0 and Q_2 to 1. Without feedback, the states of the flip-flops remain unchanged until the next pulse arrives. However, with the feedback shown, the change in Q_2 is fed back to FF1, changing its state back to 1 giving a binary reading of $Q_0 = 0$, $Q_1 = 1$ and $Q_2 = 1$. Pulse 5 then changes all states to logic 1 and pulse 6 resets the flip-flops to logic 0. Thus, we have a count of up to 6 pulses (0 to 5), or a divide-by-6 counter. It can be seen from Table 5.3 that the effect of the feedback is to skip one step in the count, i.e. $Q_0 = 1$, $Q_1 = 0$, $Q_2 = 1$. In general a feedback loop reduces the division factor of the flip-flops inside the loop by 1. In the above example (Fig. 5.4), the flip-flops inside the feedback loop are FF1 and FF2. Without feedback they divide by a factor of $2^2 = 4$. With feedback, FF1 and FF2 form a feedback block with a division factor of $4 - 1 = 3$, as shown in Fig. 5.5. With FF1 outside the feedback block, the total division factor of the counter is $2 \times 3 = 6$.

The decade counter

Figure 5.6 shows a decade or decimal counter employing two feedback loops. Two feedback blocks are thus formed: block 1 enclosing FF2 and FF3 and dividing by $4 - 1 = 3$; and block 2 enclosing feedback block 1 as well as FF1. Without feedback, block 2 would divide by $2 \times 3 = 6$. However, the feedback loop reduces the division factor by 1 (Fig. 5.6(*b*)). This gives a division factor of $6 - 1 = 5$. With FF0 outside feedback block 2, the total division factor is $2 \times 5 = 10$.

Feedback may be applied to the clear inputs of J–K bistables, as shown in Fig. 5.7. The count continues normally up to the tenth pulse when both Q_1 and Q_3 go to logic 1. The output of the NAND gate becomes 0 and all flip-flops are cleared (i.e. reset back to 0).

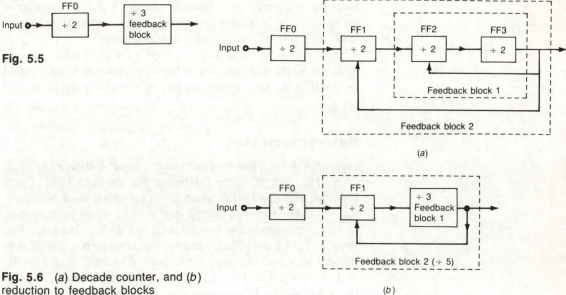

Fig. 5.5

Fig. 5.6 (*a*) Decade counter, and (*b*) reduction to feedback blocks

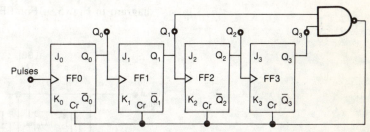

Fig. 5.7 Decade counter ($J = K = 1$)

Synchronous counters

In asynchronous counters a flip-flop in a chain will respond only when the preceding stage has completed its transition. The clock pulse ripples through the chain of flip-flops. A delay known as the carry propagation delay is therefore introduced which may be longer than the interval between the input pulses, depending on the length of the chain. In such a case, it will not be possible to read the counter between the pulses.

The operation of the asynchronous counter may be modified so that all flip-flops are clocked simultaneously by the input pulses. This mode of operation, known as the **synchronous mode**, considerably reduces the propagation delay time of the counter and hence improves the frequency of its operation. The maximum repetition rate (frequency) of the input pulses is limited by the propagation delay of the counter introduced by each flip-flop as well as by any control gates in the system. Typically, the maximum permitted input frequency of a four-element synchronous counter using TTL is 32 MHz, which is about twice that of a similar asynchronous counter. A second advantage of the synchronous type is the absence of decoding spikes at the outputs since all flip-flops change states simultaneously.

There are two possible ways of carrying the input pulses forward to enable all the flip-flops, thus allowing simultaneous triggering of the flip-flops in a synchronous counter: the series carry and the parallel carry.

Series or ripple carry

Figure 5.8 shows a four-bit synchronous counter with series or ripple carry. The input pulses are fed directly into the clock input of each flip-flop. Flip-flop FF0 is connected in the toggle mode with $J_0 = K_0 = 1$ and as such will change state at every active pulse edge t_0, t_1, t_2, t_3, etc., producing waveform Q_0 in Fig. 5.9. The other flip-flops, FF1, FF2 and FF3, are enabled and therefore may change state (toggle) with each incoming active pulse edge only when $J_1 = K_1 = 1$, $J_2 = K_2 = 1$, $J_3 = K_3 = 1$, respectively. Since $J_1 = K_1 = Q_0$, it follows that FF1 changes state every time $Q_0 = 1$, i.e. every second active pulse edge t_1, t_3, t_5, and so on, as shown on the timing diagram in Fig. 5.9. For FF2, two conditions have to be satisfied

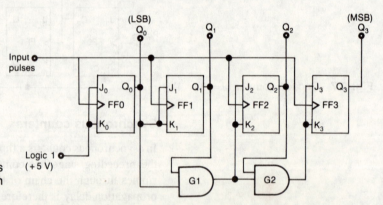

Fig. 5.8 A four-bit synchronous counter with series or ripple-through carry

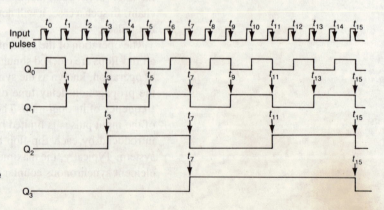

Fig. 5.9 Timing diagram for the synchronous counter in Fig. 5.8

to make $J_1 = K_1 = 1$, thus enabling the flip-flop to change its state with the incoming active pulse edge, namely $Q_0 = 1$ and $Q_1 = 1$. This occurs at t_3, t_7, t_{11} and t_{15}. For FF3, three conditions have to be satisfied: $Q_0 = 1$, $Q_1 = 1$ and $Q_2 = 1$ (to produce a logic 1 from G1 and G2). This occurs every eighth pulse, i.e. at t_7 and t_{15}.

The propagation delay of this type of counter is reduced to that introduced by the slowest flip-flop plus the delay caused by the gating circuit.

Parallel carry

Propagation delay may be reduced further, hence increasing the maximum frequency of operation of a counter chain, by the use of the parallel carry technique. The series carry suffers from the fact that although the flip-flops are triggered simultaneously the enabling signal continues to ripple through the gates, introducing considerable delay. Figure 5.10 shows a synchronous counter in which the carry

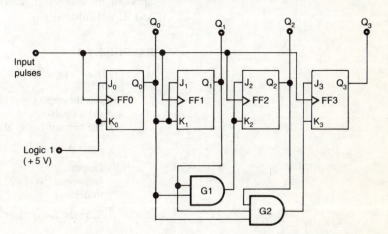

Fig. 5.10 Four-bit synchronous counter with parallel carry

is applied in parallel to all flip-flops. The principle of operation of this type of carry is the same as for the series carry. However, in this case, gate G2 does not have to wait for an output from the preceding gate G1, thus reducing the delay time due to the gates by half.

The disadvantages of a parallel-carry counter are the high fan-in requirements of the gates (two inputs for the gate feeding FF2, three for the gate feeding FF3, and so on for longer chains) and the heavy loading of the flip-flops at the beginning of the chain. (For a four-bit counter, fan-out for FF0 is $4 - 1 = 3$, and so on for longer chains.)

Commercial counters

Counters, both TTL and CMOS, are available in IC packages with programming facilities. The usual considerations apply when choosing between TTL and CMOS, namely speed, power dissipation and packaging. CMOS is slower than TTL, i.e. it has a lower operational frequency. On the other hand it has a very low power consumption and a very high component density. The following are some popular examples:

TTL	7493	Four-bit binary counter
	7490	Decade counter
	74193	Programmable counter
CMOS	4042	Seven-stage ripple-through counter
	4029	Programmable counter
	4017	Decade counter/divider

Applications of counters

Counters may be used for direct counting, division by a number, measurement of frequency, time, distance and speed, computer applications, waveform generation, and conversion between analogue and digital information.

EXERCISE 5.1

(a) Draw a logic block diagram of a counter which will give a count of 16.

(b) State the output signal of each stage after an input of (i) seven pulses; and (ii) 13 pulses.

(c) The counter in (a) above gives the following outputs:

Input pulse 0 1 2 3 4 5 6 7 8 9 10 11 12 13 14 15

Output
(converted 0 1 2 3 4 5 6 7 0 1 2 3 4 5 6 7
to denary)

State the faulty stage.

EXERCISE 5.2

(a) State one advantage and one disadvantage of synchronous counters compared with asynchronous counters.

(b) Refer to Fig. 5.11.

(i) State whether the counter shown is synchronous or asynchronous.

(ii) Assuming that the values of Q_a, Q_b and Q_c are 1, 1 and 0, respectively, list the values of Q_a, Q_b and Q_c immediately following the next seven pulses.

EXERCISE 5.3

Refer to Fig. 5.12.

(a) State the type of counter shown.

(b) State the purpose of input X.

(c) Enter the states of the outputs A, B and C after each clock pulse in the following table.

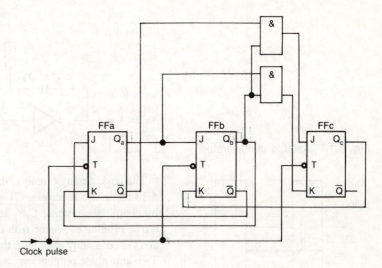

Fig. 5.11

Clock pulse

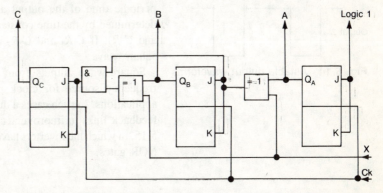

Fig. 5.12

Clock		X = 0			X = 1	
pulse	C	B	A	C	B	A
0	0	0	0	0	0	0
1						
2						
3						
4						
5						
6						
7						
8						

Timing devices

The basic element in any timing device is the oscillator. There are two basic requirements for sustained oscillation: positive feedback between the output and the input and a gain round the feedback loop of more than one.

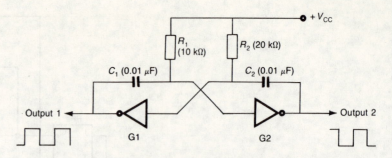

Fig. 5.13 Astable multivibrator

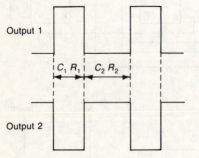

Fig. 5.14 Astable output waveforms

The basic timing circuit is the astable multivibrator shown in Fig. 5.13 using two inverters, G1 and G2. Positive feedback is provided by coupling elements, C_1R_1 and C_2R_2. The astable multivibrator is a free-running oscillator with the state of the outputs of inverters G1 and G2 constantly changing, thus producing a square or pulse output. Two anti-phase outputs may be obtained as shown in Fig. 5.14. The periodic time of the output as well as its mark-to-space ratio are determined by the time constant of the two feedback networks C_1R_1 and C_2R_2. If C_1R_1 and C_2R_2 are made equal, the output will be a square wave.

Square-wave outputs from astable multivibrators are used as a frequency source for clock pulses in digital and microprocessor applications. In such cases a quartz crystal is connected in one of the feedback links to improve accuracy and stability, as shown in Fig. 5.15, in which the inverters have been substituted by two cross-coupled NOR gates.

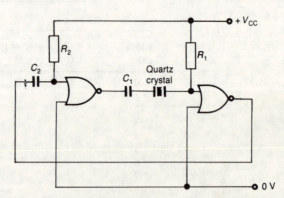

Fig. 5.15 Crystal-controlled astable multivibrator

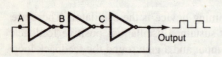

Fig. 5.16 Astable multivibrator employing propagation delay

Square-wave oscillation may also be obtained by the three-stage multivibrator circuit shown in Fig. 5.16. Provided the gain around the loop is greater than one, oscillation will take place at a frequency determined by the propagation delay introduced by the inverters. Each inverter stage introduces 180° phase shift. After the second stage, point C, the phase shift is 360°, i.e. the waveform at C is in phase with that at A. A third stage of inversion provides a further 180° phase shift, making the output out of phase with the input at A. For oscillation to occur, the waveform at the output must be in phase with input A. In other words, a delay of 180° (or one half cycle) is necessary. Such a delay does occur due to the propagation delay introduced by the inverters. Assuming that the inverters are TTL devices with a propagation delay of 10 ns each, then at the end of the chain, a 30 ns delay is introduced which provides the 180° delay necessary for oscillation. The frequency of oscillation may therefore be calculated as follows.

Total propagation delay = 30 ns.

Since total propagation delay is equivalent to 180° or one half cycle of the output, it follows that

Time for one cycle (period) = 60 ns.

$$\text{Frequency} = \frac{1}{\text{period}} = \frac{1}{60 \text{ ns}} = 16.667 \text{ MHz}.$$

The frequency of oscillation may be reduced by introducing further delay through the introduction of time-constant elements in the feedback loop, as shown in Fig. 5.17. G1 and G2 form a two-stage astable multivibrator, $C_1 R_1$ is the time-constant network and G3 is a buffer.

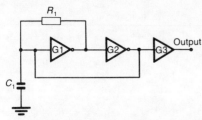

Fig. 5.17 A two-stage astable multivibrator employing a time-constant element to reduce frequency

Some TTL IC pin connections

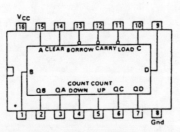

74192 Up/Down decade counter

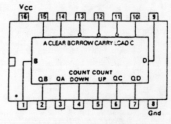

74193 Up/Down binary counter with preset inputs

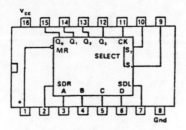

74194 A 4-bit bidirectional universal shift register

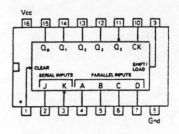

74195 4-bit parallel-access shift register

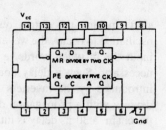

74196 4-stage presettable
ripple counter

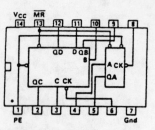

74197 Presettable binary
ripple counter

6 Memory devices

Digital systems invariably require memory facilities for permanent or temporary storage of data to perform their function. Memory devices include memory chips, magnetic discs and the ordinary cassette tape recorder.

Memory chips

A memory chip consists of a number of memory cells into which data bits may be stored (or written). The stored data may then be retrieved (or read) from the device. These memory cells are grouped together to form a memory location (e.g. one-bit, two-bit, four-bit or eight-bit memory locations). Data stored in these locations are known as **words**. A word is a group of binary bits forming the basic unit of information of a system. A four-bit word (or number) is known as a **nibble** while an eight-bit word (or number) is referred to as a **byte**. Each location is given a unique binary code known as an **address** for the purposes of identification.

Consider Fig. 6.1 which represents the arrangement for eight four-bit memory locations. Each location has four memory cells; D_0, D_1, D_2 and D_3, into which a four-bit word may be stored. Each location is identified by a unique three-digit (A_0, A_1, A_2) binary address; 000, 001, 010, ..., 110, 111. A three-digit binary number can address up to $2^3 = 8$ locations. In order to be able to address a bigger memory store with a larger number of locations, a binary address of a higher order must be used. For example: to address 16 ($= 2^4$) different locations, four digits (A_0, A_1, A_2, A_3) are necessary; to address 32 locations, five-digit addresses are required; and so on. A memory store with $2^{10} = 1024$ locations is known as having a **1 k memory**. Such a memory device requires a 10-digit address: A_0, A_1, ..., A_9. The actual size of the memory, known as its capacity, is determined by the number of bits in each location as well as the number of locations available on the device. Hence a memory chip with 1024 locations with each location consisting of two memory cells or bits has a total memory capacity of 1 k $\times$ 2 = 1024 $\times$ 2 = 2048 bits in total. Similarly, a memory chip with $2^{11} = 2048$ or 2 k locations

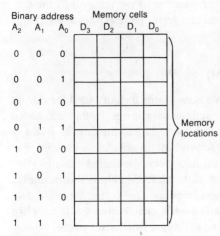

Binary address   Memory cells
A_2 A_1 A_0 D_3 D_2 D_1 D_0

Memory locations

Fig. 6.1

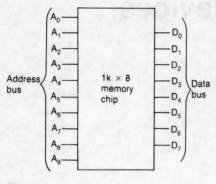

Fig. 6.2

and four bits per location has a capacity of $2 \text{ k} \times 4 = 2048 \times 4 = 8192$ bits, and so on.

Figure 6.2 shows some of the pin connections for a typical $1 \text{ k} \times 8$ bit memory chip. The chip has $2^{10} = 1024$ or 1 k locations and hence 10 address pin connections or lines: A_0-A_9. These lines are grouped together to form an **address bus**. By placing an address on the address bus, any one of the 1024 locations may be selected. For example, if A_0, A_1, ..., A_8, A_9 are set to logic levels 0, 0, 1, 0, 1, 1, 0, 1, 0 and 1, then the location with address 0010110101 in binary or 0B5 in hexadecimal is selected. For simplicity, addresses are normally stated in hexadecimal. Once a memory location is chosen, pins D_0 to D_7 provide access to the eight memory cells in that location. Lines D_0, D_1, D_2, ..., D_7 are grouped together to form a **data bus**.

Access time

Apart from the number of locations and the width of each location, memory chips have another property that must be taken into account when used in digital systems, namely access time. Access time is the speed with which a location within the memory chip can be made available to the data bus. It is the time interval between the instant that an address is sent to the memory chip and the instant that the data stored into the location appears on the data bus. For memory chips access times vary between 200 and 450 ns.

Volatile and non-volatile memories

There are two main types of memory devices: volatile and non-volatile. A volatile memory loses its stored data when the d.c. power is removed. The non-volatile type on the other hand retains the data stored inside it permanently irrespective of the removal of the d.c. power. Within these two broad categories there are several types of memory packages.

Read-only memory (ROM)

Read-only memory (ROM) is a non-volatile memory used for storing data permanently. The data stored can only be read by the user, hence its name, and no new data can be written into the device. ROM is programmed by the manufacturer in accordance with the user specifications. Once entered the data cannot be altered subsequently. The pin connection for a $1 \text{k} \times$ eight-bit ROM chip is shown in Fig. 6.3. A_0-A_9 are the ten address lines and D_0-D_7 are the eight data lines. $\overline{\text{OE}}$ is **output** (i.e. **read**) **enable**, active low; $\overline{\text{CS}}$ is the **chip select** and NC is **no connection**. ROMs are fabricated by either bipolar or MOS technology. In both cases the basic storage element is a unidirectional switch in the form of a diode or a transistor. The

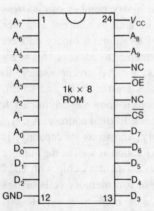

Fig. 6.3 Pin connections for a $1\text{k} \times$ eight-bit ROM chip

physical presence of the unidirectional switch in a given memory cell indicates a logic 1, and its absence indicates a logic 0.

Programmable read-only memory (PROM)

The fact that ROMs are programmed by the manufacturer means that they can be expensive unless they are produced in large quantities. Furthermore, subsequent changes to the program once it has been written into ROM are very costly. To avoid this, programmable read-only memory chips are used. PROMs fulfil the same basic function as a ROM except that they may be programmed by the user as well as by the manufacturer. The manufacturer supplies 'blank' PROM packages with all bits at logic 1 or logic 0. The user then programs the chip by '**blowing**' the unidirectional switches in selected places, thus changing the logic levels at these locations. Once programmed, PROMs cannot be altered.

Erasable programmable read-only memory (EPROM)

The main disadvantage of PROM devices is the fact they cannot be reprogrammed. Mistakes in programming thus cannot be corrected. The EPROM overcomes this by allowing the user to delete or erase the stored data and thus change the program. The stored program in an EPROM may be erased by exposing the memory cells to ultraviolet light through a 'window' on the IC package. This process takes 20 to 30 minutes at the end of which the IC is in a 'blank' state ready to be reprogrammed. EPROMs have two main disadvantages.

1. The entire memory must be erased before reprogramming. This means that selected changes cannot be made.
2. The IC has to be removed from the circuit for ultraviolet erasure. This means that the process of reprogramming cannot take place in circuit.

Electrically erasable programmable read-only memory (EEPROM)

Electrically erasable programmable read-only memory, also known as electrically alterable programmable read-only memory (EAPROM) overcomes the disadvantages of the ordinary EPROM. EEPROMs can be programmed and erased in circuit by the application of suitable electrical signals. Furthermore, individual locations may be erased and programmed without interfering with the rest of the data pattern.

Programming the EPROM

The method used for programming an EPROM varies from type to type. A typical 1k × eight-bit EPROM is shown in Fig. 6.4. The following are typical programming steps.

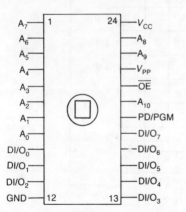

Fig. 6.4 EPROM pin connections

MATTHEW BOULTON
COLLEGE LIBRARY

1. Apply 25 V to V_{PP}.
2. Connect $\overline{OE}$ to logic 1, thus disabling the outputs.
3. Apply the address of the location to be programmed to the address inputs of the chip, $A_0 - A_9$.
4. Use the data input/output pins, $DI/O_0 - DI/O_7$ as inputs. Apply the data bits to determine which bits are to be programmed as 1 and which as 0.
5. Apply a 50 ms pulse to the PD/PGM input.
6. Repeat for all other locations.

Random access memory (RAM)

Random access memory (RAM) is a volatile memory chip which the user may read from and write into, hence it is also known as read/write memory. Locations may be accessed at random by placing the address of the selected location on to the address lines. RAMs are divided into two major categories according to the type of storage technique used; these are dynamic and static. Dynamic RAM (DRAM) stores information in the form of a charge on a capacitor. However, due to leakage, the charge is lost and has to be restored, a process known as '**refreshing**' the cell. Dynamic RAMs have the advantage of higher density and lower power consumption. Static RAM devices employ flip-flops as the basic cell and hence require no refreshing. Static RAMs will hold data as long as d.c. power is applied to the device.

Figure 6.5 shows a block diagram for a random access memory

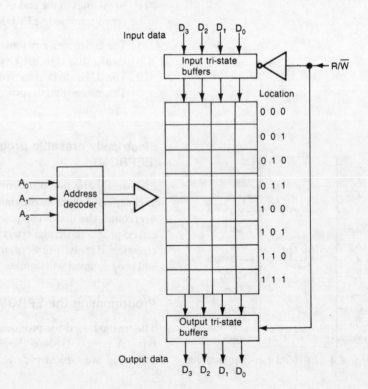

Fig. 6.5 RAM block diagram

device with three address lines, A_0-A_2, giving eight locations of one four-bit word each. Each location may be addressed separately by applying the appropriate address on the address bus. Each time a location is addressed, the $R/\overline{W}$ (read/not write) is set to logic 1 for read (active high) or is set to logic 0 for write (active low). When the $R/\overline{W}$ is set high, the output buffers are enabled and the input buffers are disabled (tri-state), allowing the contents of the selected location to appear at the output. Conversely, when the $R/\overline{W}$ line is set to logic 0, the input buffers are enabled while the output buffers are disabled allowing data to be written into the selected location.

Non-volatile RAM (NV-RAM)

One method of preserving the stored data in a normal RAM is to employ back-up batteries to maintain the d.c. supply voltage to the chip when the main supply is removed. To avoid the use of back-up batteries, the non-volatile RAM (NV-RAM) is available in which each cell has a shadow non-volatile storage transistor. The data are entered into the cells as normal but they can be transferred to the non-volatile storage cells when an enable signal is used. The main disadvantage of this type of device is its low component density, as it requires about five times the chip area of a normal RAM for the same storage capacity.

EXERCISE 6.1

A 2 k RAM has an eight-bit word in each location. State:

(a) the number of address input pins;   //
(b) the number of data pins;   8
(c) the exact number of bits that may be stored; and   16 k
(d) two essential pins other than data, address, power and ground.

EXERCISE 6.2

Refer to Fig. 6.6.

(a) State the type of memory chip shown.
(b) Calculate the storage capacity of the chip stating the number of locations, the word length and the total number of bits.
(c) State the meaning of the bar over the signal on pin 8.
(d) State the purpose of the signals present at pins 8 and 10.

Magnetic storage devices

Memory chips provide units of small storage capacity (a few kbytes). For mass data storage devices, i.e. devices that can store up to 10 Mbytes or more, one must turn to magnetic type memory stores. The increase in capacity is provided at the cost of high access times and low data transfer rates. **Data transfer rate** is defined as the number of data bits that are transferred in or out of the store per second.

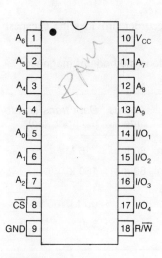

Fig. 6.6

The most popular forms of mass data storage devices are the floppy and hard discs. The **floppy disc** consists of a thin flexible sheet of plastic coated on both sides with a magnetic oxide layer. Data is recorded on a series of concentric tracks by magnetising the oxide layer using a head similar to that used in domestic cassette tape recorders. The reading/writing head is mounted on a movable arm so that it may be positioned on any track on the disc. The disc rotates under the head at a speed of 300 revolutions per minute. Each track is divided into a number of blocks of data known as sectors. Early versions of the floppy disc used an 8-inch diameter disc. Improved recording techniques made it possible to reduce the size to $5\frac{1}{4}''$ and further to $3\frac{1}{2}''$ diameter minifloppy discs. A typical $5\frac{1}{4}''$ floppy disc has a capacity of between 100 kbytes and 800 kbytes and a $3\frac{1}{2}''$ disc has a capacity of up to 1 Mbytes.

To obtain very large amounts of data storage a **hard disc** (also known as a **Winchester disc**) is used. The hard disc uses a series of metal discs mounted in a sealed unit to exclude dust (Fig. 6.7). The disc heads float just above the surface of the disc without making contact. This allows for a higher speed of rotation, an increase in the number of tracks and an expansion in the recording density along each track. Thus a $5''$ hard disc has a capacity of around 40 Mbytes with a very much higher data transfer speed of about 5Mbits per second compared with the equivalent floppy disc unit of about 160 kbits per second. Hard discs with a capacity of 320 Mbytes are also available. One disadvantage of the hard disc device is that the disc is sealed and cannot therefore be changed in the same way as a floppy can.

The magnetic tape recording system utilises a tape coated with a very thin layer of magnetic material. The tape is transported past separate reading and writing heads at a constant speed. The speed of the tape determines the data transfer rate of the device as well as its recording density. Multiple tracks are often used to improve the capacity of the tape and its data transfer.

The domestic audio cassette tape recorder provides the cheapest

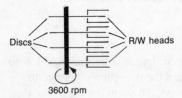

Discs — R/W heads

3600 rpm

Fig. 6.7

Table 6.1 Comparison between various methods of magnetic recording

Type	Capacity (bytes)	Speed	Access time	Data transfer rate (bits per sec)
Hard disc	20–40	3600 rev/min	30 ms	5 M
Floppy	500 k–1 M	330 rev/min	80 ms	160 k
Tape	700 k	Up to 125 in/s	—	up to 100 k
Domestic cassette	80 k	1.875 in/s (4.75 cm/s)	—	300

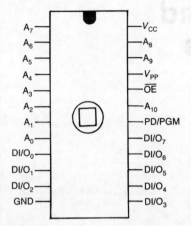

Fig. 6.8

method of mass data storage. However it suffers from low storage capacity and very slow data transfer speed of 300 bits per second. The audio cassette is designed to process speech and music rather than digital waveforms and as such cannot handle digital data directly. To overcome this, digital signals are converted to two tones with logic 0 represented by 1200 Hz and logic 1 by 2400 Hz.

EXERCISE 6.3

Refer to Fig. 6.8.

(a) State the type of memory device shown.
(b) State the storage capacity of the chip.
(c) Explain the purpose of the pin marked V_{PP}.
(d) Explain the purpose of the pin marked PD/PGM.

EXERCISE 6.4

Refer to Fig. 6.9 which shows the pin configuration of a random access memory (RAM). The hexadecimal code E1 is being written into the hexadecimal address B01. List all pin numbers which are shown to have incorrect logic levels applied to them.

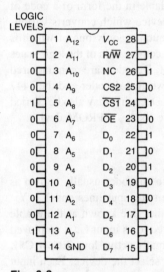

Fig. 6.9

7 Decoders and multiplexers

Decoders

In digital systems, information is available in the form of a code of one kind or another. A decoder is a device which converts various coded inputs into mutually exclusive outputs, only one of which is active at any one time. It interprets each combination of the logic states of the inputs into the appropriate output. Decoders are manufactured as MSI chips such as the 74155 dual two-line decoder and the 7447 BCD-to-seven segment decoder driver. Decoding may also be carried out by programming a ROM or, more likely, a PROM device.

The n-line to m-line decoder

Figure 7.1 shows a two-line to four-line decoder usually known as two-to-four decoder. I_0 and I_1 are the two input lines and Y_0, Y_1, Y_2 and Y_3 are the four outputs. A further line known as the enable (EN) or chip select (CS) usually referred to as input G, is employed to 'activate' the device. Enable is normally active low ($\overline{EN}$ or $\overline{CS}$), i.e. a logic 0 is necessary to enable or select the device. Each input can assume a logic 0 or a logic 1. This gives a $2^2 = 4$ possible combinations of I_0 and I_1 namely 00, 01, 10 and 11. Each of these combinations activates one and only one of the four outputs as shown in Table 7.1

It can be seen from the truth table that the input combination is the binary form of the prefix number of the output line. For example to select Y_2 (output line number 2), then the inputs must assume the combination 10 which is the binary form of 2. Similarly, for Y_3, the input combination must be 11, the binary equivalent of 3, and so on. This is the easiest way to remember the truth table for this or other similar decoders. When the $\overline{Enable}$, $\overline{EN}$ is high, no output is selected regardless of the state of the inputs.

Figure 7.2 shows the pin connection of a dual 2–4 decoder. It consists of two separate 2–4 decoders; decoder 1 consists of inputs 1A and 1B and outputs $1Y_0$, $1Y_1$, $1Y_2$ and $1Y_3$ with 1G its enable

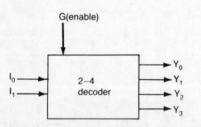

G(enable)

I_0 ⟶
I_1 ⟶

2–4 decoder

⟶ Y_0
⟶ Y_1
⟶ Y_2
⟶ Y_3

Fig. 7.1 Two-line–four-line decoder

Table 7.1

Enable ($\overline{EN}$)	Inputs I_0	I_1	Outputs Y_0	Y_1	Y_2	Y_3
0	0	0	0	1	1	1
0	0	1	1	0	1	1
0	1	0	1	1	0	1
0	1	1	1	1	1	0
1	X	X	1	1	1	1

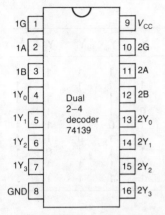

Fig. 7.2 Pin connections for a 2–4 decoder

control pin. The pins on the other side of the chip provide the inputs, output and enable connections for decoder 2.

A decoder with three inputs which provides $2^3 = 8$ outputs is known as a 3–8 decoder. A decoder with four input lines has $4^2 = 16$ output lines (4–16 decoder). Table 7.2 shows a typical truth table for a 3–8 decoder.

Address decoders

Where more than one memory chip is used in a system, it is necessary to avoid overlap of memory locations by allocating a particular memory area for each memory device. Such a task is carried out by an n-line to m-line decoder known as the address decoder, as shown in Fig. 7.3. The system in Fig. 7.3 employs eight memory chips ($M_0–M_7$) with 1k = 1024 locations each. As explained earlier, each chip requires 10 addressing lines ($A_0–A_9$) which are used to select any one of the 1024 locations within it. Address lines $A_0–A_9$ cannot however distinguish one memory chip from another. To do this, a 3-to-8 decoder is used. Address lines A_{10}, A_{11} and A_{12} are fed into the decoder as shown, with its outputs $Y_0–Y_7$ connected to the appropriate memory chip. For example, an input of 000 ($A_{10} = A_{11} = A_{12} = 0$) will select memory chip M_0 and disable all the others.

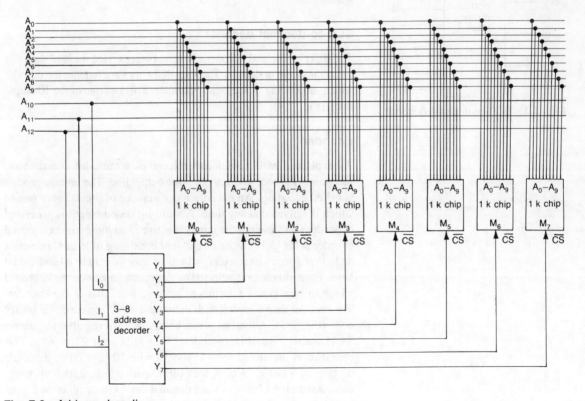

Fig. 7.3 Address decoding

Table 7.2

EN	I_0	I_1	I_2	Y_0	Y_1	Y_2	Y_3	Y_4	Y_5	Y_6	Y_7
0	0	0	0	0	1	1	1	1	1	1	1
0	0	0	1	1	0	1	1	1	1	1	1
0	0	1	0	1	1	0	1	1	1	1	1
0	0	1	1	1	1	1	0	1	1	1	1
0	1	0	0	1	1	1	1	0	1	1	1
0	1	0	1	1	1	1	1	1	0	1	1
0	1	1	0	1	1	1	1	1	1	0	1
0	1	1	1	1	1	1	1	1	1	1	0
1	X	X	X	1	1	1	1	1	1	1	1

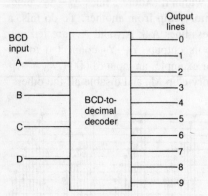

Fig. 7.4 BCD-to-decimal decoder

Similarly, an input of 011 ($A_{10} = 1$, $A_{11} = 1$ and $A_{12} = 0$) enables M_3 and so on.

Decoding may be carried out by programming a memory chip, usually a PROM. Software decoding is more flexible than the hardware type in that it may be used to select memory chips of differing capacity. It may also be easily modified to accommodate changes in the system without the need to make changes to the internal wiring.

BCD-to-decimal decoder

The BCD-to-decimal decoder converts binary coded decimal data into their equivalent in denary. The four BCD lines are fed into the decoder which, depending on the binary number, activates one of the 10 output lines (Fig. 7.4).

Encoder

As explained earlier, a decoder receives a multi-bit coded input, recognises the code and activates one output line. The inverse process is called encoding. An encoder has a number of inputs, only one of which is activated at any time. A multi-bit coded output is generated depending on which of the inputs is active. The most commonly used encoder is the alphanumeric keyboard consisting of letters, numerals and other special characters, a total of approximately 84 individual keys. Each character is selected by depressing a key on the keyboard which in turn closes a switch to activate one input to the encoder. The encoder then converts each individual input into a specific binary code (Fig. 7.5). A total of seven binary bits are required to convert 84 characters into binary codes ($2^7 = 128$, but $2^6 = 64$). This operation is illustrated for a keyboard with 10 keys only, numerals 0, 1, ..., 9 in Fig. 7.6. A four-bit output code is sufficient in this case. Assuming a binary-coded decimal output the truth table for the encoder is shown in Table 7.3. For instance, if numeral 5 is selected,

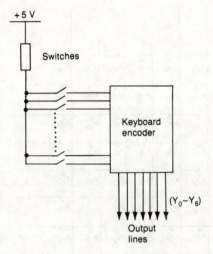

Fig. 7.5   Keyboard encoder

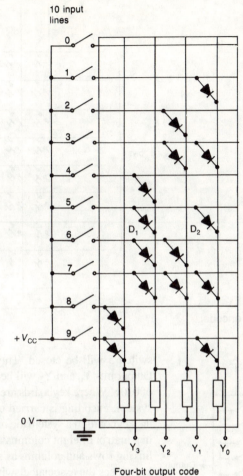

Fig. 7.6   Decimal-to-binary encoder

Table 7.3

			Number to be encoded									Outputs		
0	1	2	3	4	5	6	7	8	9	Y_3	Y_2	Y_1	Y_0
0	—	—	—	—	—	—	—	—	—	0	0	0	0
—	1	—	—	—	—	—	—	—	—	0	0	0	1
—	—	2	—	—	—	—	—	—	—	0	0	1	0
—	—	—	3	—	—	—	—	—	—	0	0	1	1
—	—	—	—	4	—	—	—	—	—	0	1	0	0
—	—	—	—	—	5	—	—	—	—	0	1	0	1
—	—	—	—	—	—	6	—	—	—	0	1	1	0
—	—	—	—	—	—	—	7	—	—	0	1	1	1
—	—	—	—	—	—	—	—	8	—	1	0	0	0
—	—	—	—	—	—	—	—	—	9	1	0	0	1

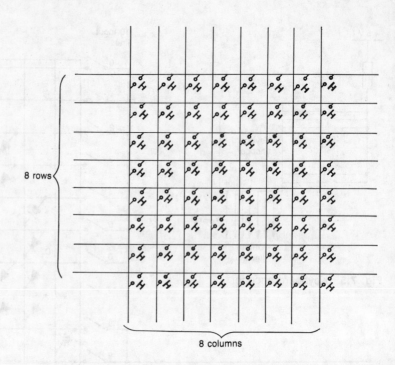

Fig. 7.7 8 × 8 matrix encoder

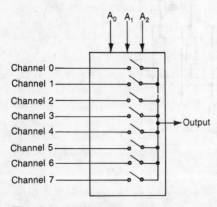

Fig. 7.8 Multiplexing

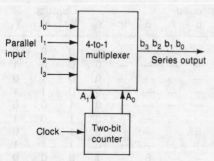

Fig. 7.9 Multiplexing used as a serial-to-parallel converter

switch 5 will be closed. This will forward bias diodes D_1 and D_2. Output bits Y_0 and Y_2 will be pulled to logic 1 giving a binary output of 0101. Where keyboards are used as peripheral devices in a computer system, encoding is carried out by the microprocessor interrogating the keyboard in a systematic manner. In this case the keys are arranged in a matrix of eight columns and eight rows with individual switches linking rows and columns as shown in Fig. 7.7. The operation of any key closes the associated switch, selecting a row and a column. The microprocessor scans the columns addressing each one in turn in a process known as **polling**. When a closed switch is detected, the appropriate code is then read by the microprocessor.

Multiplexer

A multiplexer is an electronic device that performs the function of a very fast rotary switch. It connects several input channels, one at a time to one common output line. The input channels are therefore made to share a single communication line with each channel occupying the line for a certain amount of time. This **time-sharing** technique is known as time multiplexing. The input channels may be selected in a prearranged sequence or by the application of the appropriate selection of a channel address as shown in Fig. 7.8. Address lines A_0, A_1 and A_2 form a three-bit channel select capable of selecting any one of eight input channels: channels 0−7.

Multiplexers may be used as parallel-to-serial data converters as shown in Fig. 7.9. Consider a four-bit word available in parallel so

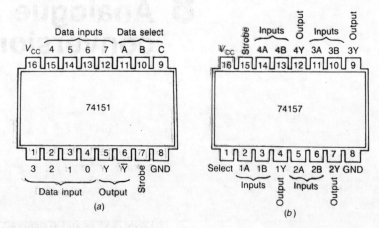

Fig. 7.10 (a) One-of-eight multiplexer, and (b) quad two-input multiplexer

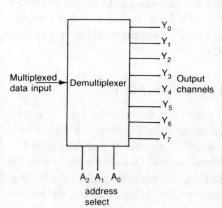

Fig. 7.11 Demultiplexing

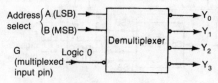

Fig. 7.12 Demultiplexer used as a decoder

that I_0 represents bit b_0, I_1 bit b_1, and so on. By means of the two-bit counter, the address select code is continuously incremented so that it is 00 for a period t_1, 01 for a second period t_2 and so on. In this way, the output of the multiplexer will be b_0 for period t_1, b_1 to t_2, and so on, rearranging the parallel input data into a serial form.

Two typical multiplexer chips are shown in Fig. 7.10. The 74151 shown in Fig. 7.10(a) is a one-of-eight multiplexer in which D_0, D_1, ..., D_7 are the input data pins. Y and its complement $\overline{Y}$ provide the output and A, B and C are the three channel-select (address) inputs. Figure 7.10(b) shows a quad two-input multiplexer. Each of the four Y outputs (1Y, 2Y, 3Y and 4Y) is connected to the corresponding A input (1Y to 1A, 2Y to 2A, and so on) when the select input is at logic 0, and to the B input (1Y to 1B, 2Y to 2B and so on) when the select input is at logic 1. The Y outputs are set to logic 0 when the **strobe** control input is set to logic 1. When the strobe is set to logic 0, the Y outputs acquire the appropriate logic level determined by the setting of the input pins.

Demultiplexer

Data channels transmitted in multiplexed form have to be separated back to their original form before they can be used. This process is known as demultiplexing. A demultiplexer is therefore a device which transmits input data coming in on a single channel on to one of several output lines, the particular line being selected by means of an address A_0–A_2 in Fig. 7.11. The demultiplexer may be used as a decoder if the data input is held at logic 0 which becomes a chip enable (Fig. 7.12).

8 Analogue and digital conversion

Logic devices and systems accept only digital signals. However, most signals such as those associated with transducers are analogue in form. Before such signals are fed into a digital system, they have to be converted into a digital form by an analogue-to-digital converter (ADC). Conversely, the output of a digital system which is in digital form invariably has to be converted into analogue form by a digital-to-analogue converter (DAC).

The analogue-to-digital converter (ADC)

The analogue-to-digital converter takes the analogue input, samples it and then converts the amplitude of each sample to a digital code as shown in Fig. 8.1. The output is a number of parallel digital bits (four in Fig. 8.1) whose simultaneous logic states represent the amplitude of each sample in turn. A variety of codes may be used for such representation, the most popular being the binary code which will be used throughout this chapter.

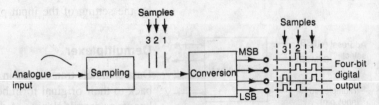

Fig. 8.1 The analogue-to-digital converter (ADC)

Sampling

For satisfactory results the analogue signal must be sampled at a rate at least twice the highest frequency of the original analogue input, as shown in Fig. 8.2. This sampling rate is known as the **Nyquist** rate. When the samples are reproduced and the dots are joined together (Fig. 8.2(*b*)), the reconstructed waveform contains all the information of the original analogue waveform.

If the sampling rate is low, i.e. comparable to the frequency of the

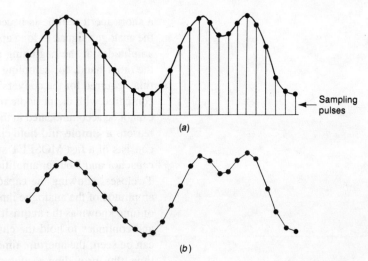

Fig. 8.2 (*a*) Sampling, and (*b*) reconstruction of original signal

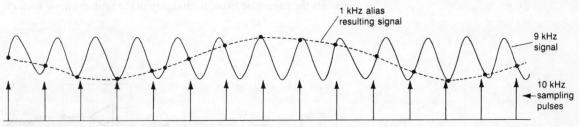

Fig. 8.3 Aliasing

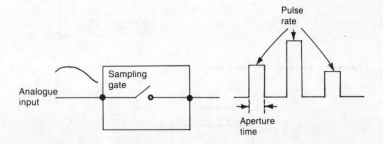

Fig. 8.4

analogue signal, a curious effect known as **aliasing** occurs. Figure 8.3 shows the effect of sampling a 9 kHz analogue signal using a sampling pulse rate of 10 kHz. Because the samples are too infrequent, they represent just one value of the signal but at a slightly different point of each cycle, resulting in a sine wave with a frequency equal to the difference of the two frequencies: 10 kHz − 9 kHz = 1 kHz, in this particular case.

Before conversion takes place the signal is fed into a sampling gate (Fig. 8.4). The gate closes for a short period of time, known as the **aperture time**, to hold the level of the sample at a steady value long enough for the converter to carry out the conversion process. Aperture time is restricted by two opposite considerations. On the one hand a short aperture time is necessary where there is a rapid change in the analogue signal. A long aperture time will result in different signal amplitudes at the beginning and at the end of a single sample. On the other hand, for sampling rates greater than 3 kHz, the aperture time required for the conversion process to be completed is too large compared with the periodic time of the sampling frequency and thus cannot be accommodated in the period between the samples. For these reasons a **sample-and-hold** circuit is employed. The circuit (Fig. 8.5) consists of a fast MOSFET switch, T_1, together with a low-leakage capacitor and a buffer amplifier. At the arrival of the sampling pulse, T_1 closes, allowing the capacitor to charge up to the instantaneous amplitude of the analogue input as shown in Fig. 8.6. After a period of time known as the **acquisition time**, the capacitor is fully charged and continues to hold the charge until the next sampling pulse. As can be seen, the aperture time is slightly longer than the acquisition time, thus providing a safety margin. The buffer amplifier is a unity gain voltage follower having a very high input impedance and therefore prevents the capacitor from discharging in the time during which the

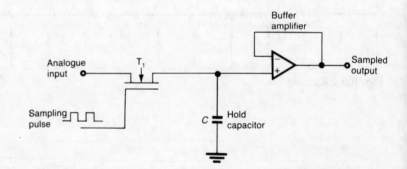

Fig. 8.5   Sample-and-hold circuit

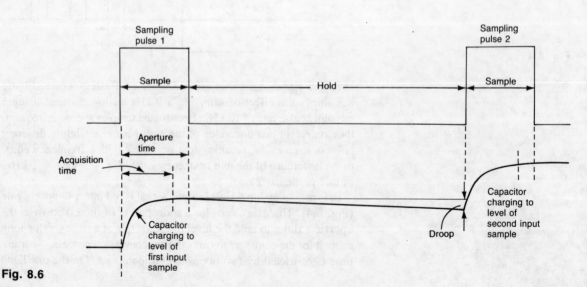

Fig. 8.6

transistor switch is open. In practice, the voltage across the capacitor falls slightly due to the leakage current of the MOS switch, the self-discharge of the capacitor through its dielectric and the input current of the buffer amplifier. This voltage drop is known as the **droop**.

Conversion

The final stage of the ADC is the conversion process itself. A number of levels, e.g. 0.25, 0.5, 0.75, 1.0, and so on, are established with each level being given a binary code. This is known as **quantising**. The number of these quantum levels is determined by the number of bits at the converter output. For instance, in a three-bit ADC, the binary output can have a coded value from 000 to 111, a total of eight levels. Now suppose that a scale or a quantum of 250 mV is used, then Table 8.1 listing sample voltages and their corresponding binary codes will apply, giving a maximum voltage of 1.75 V.

The quantum, 250 mV, represents the **resolution** of the converter which is defined as the smallest step of the input voltage that can be recognised and accurately converted into a digital output.

With the input being analogue (i.e. continuous), sample voltages will invariably fall between the quantum levels. Hence there is always an element of uncertainty or ambiguity in terms of the value of the least significant bit. This uncertainty or ambiguity is inherent in any digital coding or digital display of analogue values. It is known as the **quantising (or ± 1) error** and it is equal to $\frac{1}{2}$ quantum level. For instance, using Table 8.1 with a quantum level of 250 mV, a binary code of 110 may represent a voltage from 1.375 V to 1.625 V, i.e. quantised level $6 \pm \frac{1}{2}$ quantum $= 1.500$ V $\pm \frac{1}{2}$ 0.250 V $= 1.500 \pm 0.125$ V.

While the quantising error cannot wholly be avoided, it can be minimised by improving the resolution of the converter through increasing the number of bits used and thus reducing the quantum level and with it the quantising error. For instance if in our previous example a four-bit ADC is used, then given the same maximum voltage of 1.75 V, the new resolution is given by

$$\frac{\text{maximum output voltage}}{(2^4 - 1)} = \frac{1.75 \text{ V}}{(16 - 1)}$$

$$= \frac{1.75}{15} = 0.1167 \text{ V} = 116.7 \text{ mV}.$$

Further improvements may be obtained with an eight-bit ADC ($2^8 = 256$ levels) or a 12-bit ADC ($2^{12} = 4096$ levels).

Methods of analogue-to-digital conversion

There are several methods available for the measurement of sample voltage levels and their conversion to a binary code. These methods

Table 8.1

Level	Sample voltage (V)	Binary code		
		MSB		LSB
0	0	0	0	0
1	0.25	0	0	1
2	0.50	0	1	0
3	0.75	0	1	1
4	1.00	1	0	0
5	1.25	1	0	1
6	1.50	1	1	0
7	1.75	1	1	1

differ in the accuracy they provide and in the speed of conversion, indicated by the conversion time. **Conversion time** is the time interval between the start of the conversion process and the appearance of the binary code at the output. Converters range from the slow counter-ramp type (conversion time in milliseconds) used in digital-indicating instruments to the ultra-fast parallel or direct comparison type (conversion time in nanoseconds) used in instrumentation and control. In this chapter we will deal with two types of ADC: the successive-approximation and the counter-ramp converters.

Successive-approximation ADC

In this type of ADC the sample input is compared with successive voltages generated by the successive approximation register (SAR) programmer (Fig. 8.7). It starts at half full scale (MSB = 1) and if this is different from the voltage of the sample input, the SAR then adjusts the digital output accordingly.

The conversion thus begins with the most significant bit of the binary output. It is assigned a logic 1, applied to the digital-to-analogue converter, which converts it into an analogue value to be compared with the voltage level of the sample input. If a 1 at the MSB represents a larger value than the input, the 1 is replaced with a 0. If a 1 at the MSB represents a smaller value than the input, it is retained. The process is then repeated for the next bit and so on until the voltage output of the DAC is equal to the input level to within one half of the least significant bit. This difference is of course caused by the quantising error inherent in digital conversion.

Successive approximation may be controlled by software, hence its popularity in microprocessor-based systems. It is relatively fast and small in size. An additional advantage is that each sample is converted in the same amount of time. In other words, conversion time remains constant regardless of the level of the input and is entirely determined by the frequency of the driving clock and the resolution

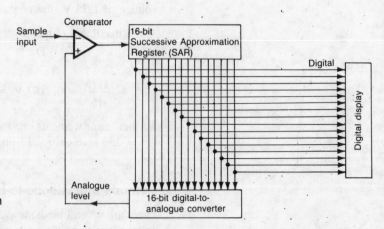

Fig. 8.7 Successive approximation converter

of the converter. For instance an eight-bit converter has to determine the logic level of each of the eight bits in sequence starting from the most significant. Given a clock frequency of 10 kHz,

$$\text{conversion time} = 8 \times \text{clock period} = 8 \times 0.1 \text{ ms} = 0.8 \text{ ms}.$$

If the clock frequency was increased to 1 MHz, the conversion time will be reduced to 8 μs.

The disadvantages of the successive approximation converter are low noise immunity and the need for a precision digital-to-analogue converter and a high performance comparator.

The counter-ramp ADC

A four-bit counter-ramp analogue-to-digital converter is shown in Fig. 8.8. The sample input is applied to the comparator while at the same time the counter starts counting upwards from 0000. The DAC converts the binary output of the counter into the ramp or staircase shown in Fig. 8.9. The maximum number of steps is determined by the number of output bits, i.e. the resolution of the ADC. Thus for a four-bit converter, 2^4 or 16 steps are available. The height of each step represents the quantum or minimum voltage increment of the converter V_{min}. Assuming $V_{min} = 100$ mV, then for the four-bit converter shown, the maximum or full scale voltage is $2^4 \times 100 = 16 \times 100 = 1600$ mV or 1.6 V. The level of the ramp is continuously compared with the input sample. While the ramp voltage is below the input sample voltage, the comparator output is positive enabling the AND gate and allowing the clock pulses to trigger the counter. When the ramp exceeds the voltage level of the input sample, the comparator output goes low, disabling the AND gate which in turn stops the counter and with it the ramp. The binary output of the counter is latched into the data register and a four-bit output is produced. The counter is then reset and the process is repeated for a second sample and so on.

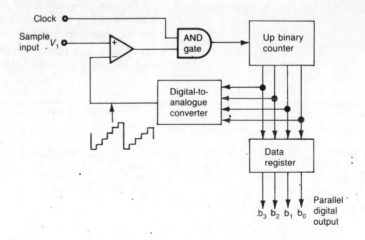

Fig. 8.8 Counter-ramp ADC

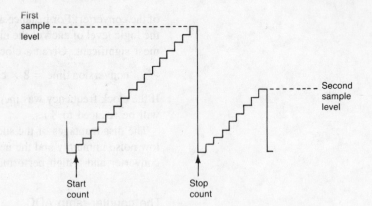

First
sample
level

Second
sample
level

Start
count

Stop
count

Fig. 8.9

A major drawback of this type of converter is its unequal conversion times. As illustrated in Fig. 8.9, if the sample level is low, conversion is completed in few clock cycles; whereas if the input level is high, conversion takes longer to complete. **Minimum conversion time** occurs when the input, V_1 is equal to the minimum voltage increment of the converter ($V_1 = V_{min}$). In this case conversion is completed in one clock period. Assuming a clock frequency of 100 kHz, then minimum conversion time = 0.01 ms or 10 μs. **Maximum conversion time** occurs when $V_1 = V_{max}$ in which case conversion time = 2^n × clock period, where n is the number of bits in the counter. Given a clock frequency of 100 kHz, then for a four-bit converter the maximum conversion time = $2^4 × 0.01 = 16 × 0.01 = 0.16$ ms or 160 μs.

The digital-to-analogue converter (DAC)

The digital-to-analogue converter receives a parallel digital input and converts it back to a voltage (or a current) value that is represented by the binary input. If this is repeated for successive digital inputs, an analogue waveform may be produced. For instance, for a three-bit binary input, eight levels are produced with 000 representing zero output and 111 representing a maximum voltage output determined by the reference voltage V_{ref} in Fig. 8.10. Other inputs are reproduced as a proportion of V_{ref}, e.g. 001 as $\frac{1}{8} V_{ref}$, 011 as $\frac{3}{8} V_{ref}$ and 101 as $\frac{5}{8} V_{ref}$. Each bit (b_2, b_1, b_0) of the binary input is

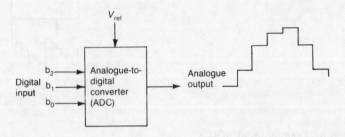

V_{ref}

Digital
input

b_2
b_1
b_0

Analogue-to-
digital
converter
(ADC)

Analogue
output

Fig. 8.10

reproduced in accordance with its weighting giving the following general formula.

$$\text{Output level} = V_{\text{ref}} (b_2/2 + b_1/4 + b_0/8).$$

The weighted-resistor ladder DAC

This type of digital-to-analogue converter uses a summing amplifier as shown in Fig. 8.11. The summing resistors at the input are chosen to give the appropriate weighting in accordance with the binary code used. The weighting or gain for each bit is given by the ratio R_4/R where R is the series resistor of the relevant bit. By making $R_0 = 8 R_3$, $R_1 = 4 R_3$ and $R_2 = 2 R_3$ the correct weighting is established. For example, using the values given in Fig. 8.11 the gain for the most significant bit (MSB), b_3 is

$$\frac{R_4}{R_3} = \frac{10}{10} = 1.0$$

and that for b_2 is

$$\frac{R_4}{R_2} = \frac{10}{20} = 0.5, \text{ and so on.}$$

S_0, S_1, S_2 and S_3 are digitally controlled switches which ensure that the series resistors are connected to $V_{\text{ref}} (-5 \text{ V})$ when its associated bit is at logic 1 and to 0 V when it is at logic 0.

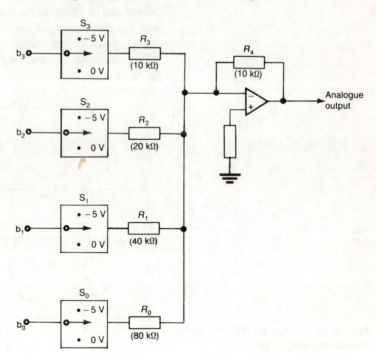

Fig. 8.11 Weighted-resistor ladder type digital-to-analogue converter

The major problem of this simple type of digital-to-analogue converter is that the range of resistor values necessary for high-resolution systems will be excessively large. For instance an eight-bit input requires a range of $2^7:1$ or 128:1, a 12-bit input needs a range of resistor values of $2^{11}:1$ or 2048:1, and so on. The accuracy and stability of this type of DAC depend primarily on the absolute accuracy of the resistors and the tracking of each resistor with temperature. Resistors must therefore have very close tolerance and high stability so that their relative values remain constant throughout. This becomes very difficult to achieve for DACs with resolutions above a few bits.

The R–$2R$ ladder DAC

This is the most commonly used type of digital-to-analogue converter. It overcomes the problem of accuracy associated with the weighted-resistor type by reducing the range of resistor values required to 2:1 (only two values of resistors are used) regardless of the resolution of the converter.

Consider the two-bit converter shown in Fig. 8.12, where S_0 and S_1 are digitally controlled switches. The ladder network is terminated with $R_t = 2R$ which ensures that at any node N_1 and N_0 the resistance looking left, right and towards the switch is $2R$. For instance with MSB = 1, S_1 switches to $-V_{ref}$ and S_0 to 0 V, giving the circuit shown in Fig. 8.13(a). The resistor network to the left of N_1 may be reduced to $2R$ giving the circuit in Fig. 8.13(b). Since the inverting input of the operational amplifier, point p, is at virtual earth then for the purposes of calculating the voltage at N_1, the circuit may be further simplified to that in Fig. 8.13(c). This gives a voltage V_{N_1} at N_1 of

$$\frac{-V_{ref} \times 2R}{2R + R} = -\frac{V_{ref}}{3}.$$

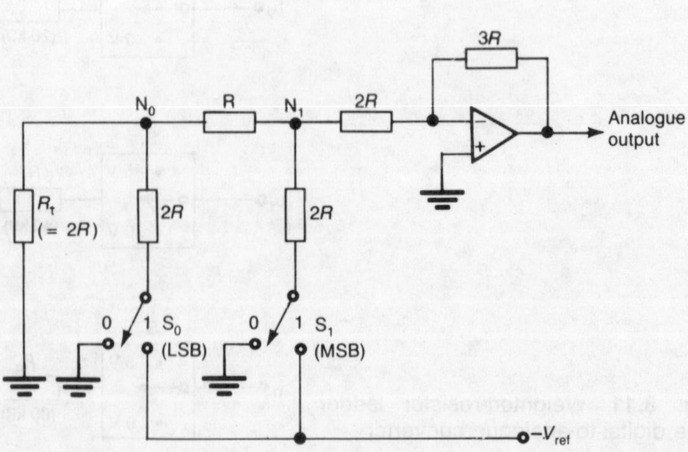

Fig. 8.12 A two-bit R–$2R$ ladder digital-to-analogue converter

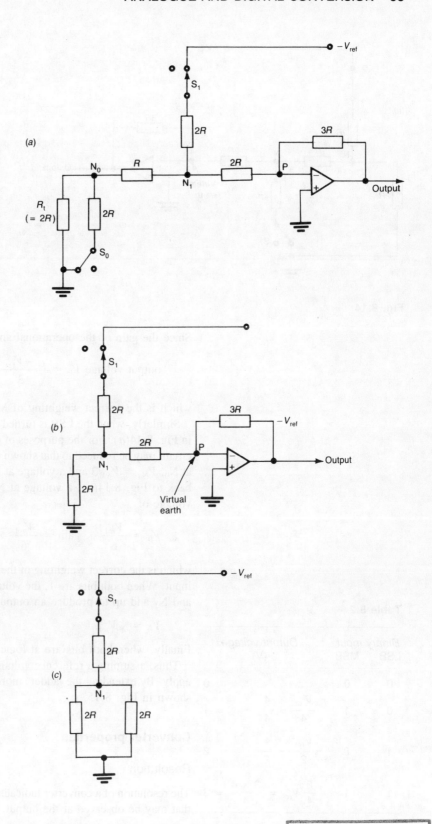

Fig. 8.13

MATTHEW BOULTON
COLLEGE LIBRARY

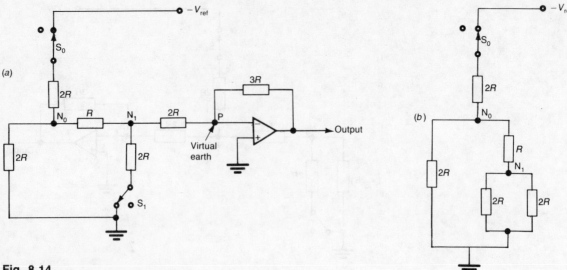

Fig. 8.14

Since the gain of the operational amplifier at N_1 is $-3R/2R$ then

$$\text{output voltage } V_0 = \left(\frac{-V_{\text{ref}}}{3}\right)\left(-\frac{3R}{2R}\right) = \frac{V_{\text{ref}}}{2}$$

which is the correct weighting of MSB of a two-bit input.

Similarly, when the LSB is turned on, the circuit will be as shown in Fig. 8.14(a). For the purposes of calculating the voltage at N_0 the circuit may be reduced to that shown in Fig. 8.14(b), giving a voltage at N_0, $V_{N_0} = V_{\text{ref}}/3$ and a voltage at N_1 of $\frac{1}{2} V_{N_0} = V_{\text{ref}}/6$. Referring back to Fig. 8.14(a), a voltage at N_1 of $V_{\text{ref}}/6$ produces an output voltage of

$$V_0 = \frac{V_{\text{ref}}}{6} \times \text{gain} = \frac{V_{\text{ref}}}{6} \times \frac{3R}{2R} = \frac{V_{\text{ref}}}{4},$$

which is the correct weighting of the least significant bit of a two-bit input. When both bits are 1, the voltages generated at both nodes N_0 and N_1 add up to produce an output of

$$V_0 = \tfrac{3}{4} V_{\text{ref}}.$$

Finally, when both bits are at logic 0, the output is also zero.

Thus, assuming a reference voltage $V_{\text{ref}} = -4$ V, Table 8.2 will apply. By extending the ladder, more bits may be accommodated as shown in Fig. 8.15.

Converter properties

Resolution

The resolution of a converter indicates the smallest change in the input that may be observed at the output. It is expressed as 1 part in 2^n,

Table 8.2

Binary input		Output voltage		
LSB	MSB	(V)		
0	0			0
0	1	$\dfrac{V_{\text{ref}}}{4} = \dfrac{4}{4}$		= 1
1	0	$\dfrac{V_{\text{ref}}}{2} = \dfrac{4}{2}$		= 2
1	1	$\dfrac{3}{4} V_{\text{ref}} = \dfrac{3 \times 4}{4}$		= 3

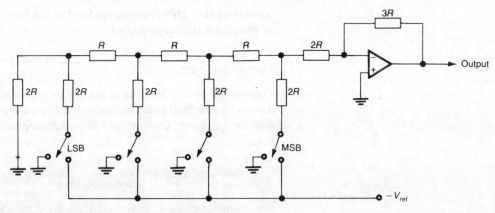

Fig. 8.15 A four-bit $R-2R$ DAC

where n is the number of bits. It is sometimes expressed as a percentage. For example, a four-bit converter has a resolution of 1 part in 2^4 or 1 part in 16. This may be expressed as

$$\tfrac{1}{16} \times 100\% = 6.25\%.$$

Higher resolution is obtained by increasing the number of bits. An eight-bit converter, for instance, has a resolution of 1 part in 2^8 or 1 part in 256 (0.39%), and so on.

Accuracy

The accuracy is the difference between the output of the converter and its true value expressed as a percentage of the maximum (full-scale) output.

$$\text{Accuracy} = \frac{\text{actual output} - \text{true output}}{\text{maximum output}} \times 100\%$$

$$= \frac{\text{Error}}{\text{Maximum output}} \times 100\%.$$

For example, given a maximum output of 10 V and an accuracy of $\pm 0.2\%$, then error $= \pm 0.2\% \times 10\ \text{V} = \pm 0.002 \times 10 = \pm 0.02\ \text{V} = \pm 20\ \text{mV}$.

Settling time

Settling time is defined as the time required for the digital output to reach its new value following a change in the input level.

Conversion time

Conversion time, also known as acquisition time, is the time required to complete the conversion of one sample level. For converters where conversion time changes with the level of the input, the worst case or maximum is always quoted.

Conversion rate

Conversion rate is sometimes quoted by manufacturers of ADC devices. It is defined as the highest rate at which the analogue samples may be converted. Conversion rate is the reciprocal of conversion time.

$$\text{Conversion rate} = \frac{1}{\text{conversion time}} \text{ (conversions per second).}$$

For example, given a conversion time of 10 μs, then

$$\text{Converstion rate} = \frac{1}{(10 \times 10^{-6})} = 0.1 \times 10^6 \text{ conversions/s}$$

$$= 100\,000 \text{ conversions/s.}$$

Bit rate

The bit rate f_b is defined as the number of bits produced per second by the converter. Given a frequency of f kHz, then an n-bit converter has a bit rate given by

$$f_b = n \text{ bits} \times f \text{ kHz} = nf \text{ kbits per second.}$$

For example, an eight-bit converter with a sampling frequency of 20 kHz has a bit rate $f_b = 8 \times 20 = 160$ kbits/s.

Bandwidth

The bandwidth of a converter is defined as the maximum frequency generated by the sequence of bits on a digital line. The maximum rate of change, i.e. maximum frequency, is obtained when the bits alternate between 1 and 0 as shown in Fig. 8.16. One cycle of the waveform in Fig. 8.16 having a periodic time T, contains two bits. It follows therefore that the frequency of the waveform is half the bit rate fb

$$\text{bandwidth} = \tfrac{1}{2}fb = \tfrac{1}{2}nf.$$

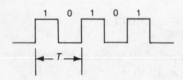

Fig. 8.16

9 Microprocessor-based systems

The microprocessor is a VLSI device that may be programmed to carry out a variety of functions and instructions. In a system, the microprocessor chip is called the central processing unit (CPU) or the microprocessor unit (MPU). The CPU consists of two main parts: the arithmetic and logical unit (ALU) and the control unit. The ALU carries out the various operations in accordance with a set of instructions known as the program. The control unit is responsible for synchronising the operation of the various units of the system including the microprocessor itself and for regulating the timing of instructions and the flow of data within the CPU as well as between the CPU and other units of the system. Microprocessors are mounted as a 40-pin dual-in-line (DIL) package shown in Fig. 9.1.

The capacity, or bit size, of a microprocessor chip is determined by the number of the data bits it can handle. A four-bit chip has a four-bit data capacity and an eight-bit chip has an eight-bit data width, and so on. The most popular type is the eight-bit microprocessor used in personal and home computers such as Zilog's Z80, Intel's 6502 and Motorola's 6809. 16-bit microprocessors such as Zilog's Z800, Intel's 8086 and Motorola's 68000 are also widely used in personal computers and industrial controllers whilst 32-bit systems are used in minicomputers. The small four-bit processors are used for dedicated minicontrollers used in domestic appliances.

The basic organisation of an eight-bit microprocessor-based system is shown in Fig. 9.2. It shows the manner in which the various elements are connected to each other in a typical system. It consists of the following units:

> central processing unit (CPU);
> memory chips (RAM and ROM);
> address decoder chip;
> input and output interface chips (PIO and UART); and
> the bus structure.

The CPU is a single chip containing all the necessary circuitry to interpret and execute program instructions in terms of data

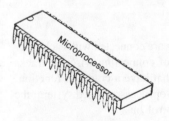

Fig. 9.1 Microprocessor chip

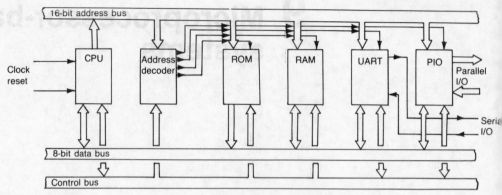

Fig. 9.2 Microprocessor-based system

manipulation, logic and arithmetic operations and timing and control of the system. RAM and ROM are the two types of memory chips that are normally used. Other types such as PROM and EPROM may also be used. The memory chips consist of a number of memory locations where data in the form of digital bits are stored. Each location can normally store an eight-bit (i.e. one byte) number or word, although two four-bit, four two-bit or eight one-bit memory chips may be used in parallel. Each location has a unique 16-bit (two-byte) address falling within the full available range of addresses, 0000 to FFFF (in hex), a total of $2^{16} = 65\,536$ or 64 k. The high byte of the address, bits A8 to A15 (the two hex digits at the left-hand side) is known as the page. For example, address 002F is on page 00 (zero page) and address 2B53 is on page 2B. It follows that there are a total of $2^8 = 256$ pages of memory, with each page containing $2^8 = 256$ locations. The input/output interface units connect the system to external devices. Two types of I/O devices are shown: the PIO, parallel input-output (also known as PIA, parallel interface adaptor) is a parallel programmable input/output interface which provides a parallel interface; and the UART (universal asynchronous receiver/transmitter) which provides a serial interface. The I/O interface units are bidirectional, providing a link to and from the system with peripheral devices such as keyboards, VDUs and transducers or drive circuitry for stepper motors, LEDs and relays. The address decoder selects the appropriate chip to be addressed by the CPU.

The bus structure

The hardware elements described above are connected with each other by a bus structure. A bus is a group of connecting wires or tracks used as paths for digital information that have a common function. There are three types of buses in a microprocessor-based system: the data bus, the address bus and the control bus.

The **data bus** is used to transfer data between the CPU and other elements in the system. Since data has to move in and out of the microprocessor, the data bus must be bidirectional, hence the bidirectional arrows.

The **address bus** is used to carry the address of memory locations to retrieve data, i.e. read, from memory devices, or to store, i.e. write, data into memory locations. It is also used to address other elements in the system such as the input/output interface units. The address bus is uni-directional, carrying 16 bits of digital information simultaneously.

The **control bus** carries all the control signals of the CPU. The number of control lines involved depends on the microprocessor used and the design of the system. The control bus provides four major functions:

memory synchronisation;
input/output synchronisation;
CPU scheduling, such as interrupt; and
other utilities such as reset and clock.

The main control signals of a CPU are given in the following sections.

The clock pulse signal

A crystal-controlled oscillator is used to provide the timing clock pulses for the microprocessor system. An example of a clock pulse generator circuit is shown in Fig. 9.3. The clock control signal Φ synchronises the movement of the data around the various elements of the system and determines the speed of operation. Clock frequencies vary between the relatively slow 6502 microprocessor at 1 MHz to the faster 16-bit 68000 chip at up to 12 MHz.

Read (RD) and write (WR)

The CPU determines the direction of data transfer to or from the microprocessor chip. This function is carried out by the read and write

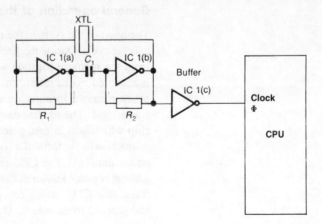

Fig. 9.3 Clock oscillator circuit

control lines. In a read operation when the CPU is receiving data from memory, the read line is active, allowing data to be transferred to the CPU. In a write operation, when the CPU is sending data to memory, the write line is active, enabling data transfer from the CPU to memory.

Interrupts

When a peripheral device such as a printer or a transducer needs attention, the main program may be interrupted temporarily by an interrupt control signal. After servicing the peripheral device, the CPU returns to the original program at the point where it was interrupted. There are several types of interrupts, e.g. interrupt request (IRQ) where the CPU will complete the current instruction that is being executed before recognising the interrupt. Halt is another type of interrupt signal which halts the main program to allow an external source or device to execute a different program.

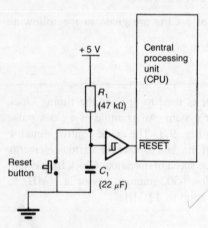

Fig. 9.4 CPU reset circuit

Reset ($\overline{\text{RES}}$ or $\overline{\text{RESET}}$)

This is a type of interrupt which overrides all other inputs, stops the CPU program and resets and starts up the microprocessor. Figure 9.4 shows a reset circuit for a microprocessor. When the push button switch is closed, capacitor C_1 is discharged and the $\overline{\text{RESET}}$ pin is taken to logic 0. All read and write operations are suspended. When the press button is released, C_1 charges up through resistor R_1 taking the $\overline{\text{RESET}}$ pin to logic 1. When that happens, the microprocessor immediately commences an initialisation sequence. This start-up sequence consists of directing the CPU to the start-up programme of the system.

All control signals are either active high, i.e. active when at logic 1 (RD for the 6502) or active low, i.e. active when at logic 0 ($\overline{\text{RD}}$ and $\overline{\text{WR}}$ for the Z80).

General operation of the system

The heart of the system, the microprocessor, operates on a fetch and execute cycle. During the fetch phase, the CPU receives the instruction from the memory location where the program is stored. The fetch is achieved by the microprocessor placing the address of the appropriate memory location on the address bus and enabling the read control line. The address decoder will select the appropriate memory chip which will in turn place the contents of that location, i.e. the instruction in the form of a coded eight-bit word known as the **opcode** on the data bus. The CPU receives the instruction, stores it into an internal register known as the instruction register. During the execute phase, the CPU having received the instruction will then decode it and proceed to execute it. This is carried out by the CPU generating the necessary timing and control signals for the execution of that particular instruction. The execute phase may involve simple arithmetic operation, e.g. add or subtract or more complex data transfer from or to memory or peripheral devices. Both the fetch and execute phases may take more than one clock cycle to complete depending on the nature of the instruction. When the instruction is completed, the microprocessor then places the next program address, i.e. the address where the next instruction is stored, on the address bus commencing another fetch and execute cycle, and so on.

Bus buffers

The bus connects the microprocessor to all memory and interface devices in the system. However, microprocessors being of the MOS type lack the drive needed for a large system. For this reason, bus drivers or buffers are used to boost the driving capability of the buses.

There are two types of drivers: the transmitter for driving the bus and the receiver for listening to the bus. In bidirectional buses such as the data bus, a transmitter/receiver is employed which is known as a **bidirectional buffer** or **transreceiver** as shown in Fig. 9.5. Tristate buffers may also be employed to disconnect the bus from the microprocessor. This techique is employed when an external device takes control of the system from the on-board CPU.

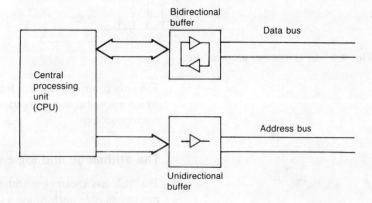

Fig. 9.5 Data and address buffers

CPU architecture

A simplified block diagram of an eight-bit microprocessor chip is shown in Fig. 9.6. In practice a more complex architecture is utilised depending on the size and make of the CPU. Connections to the other chips in the system are made via the data, address and control buses. Within the CPU itself, connections between the various elements are made via the eight-bit internal data bus and the control lines from the timing and control logic block. The internal data bus is connected to the system's data bus via a bidirectional buffer as shown. The timing and control logic provides the control signals for the whole system.

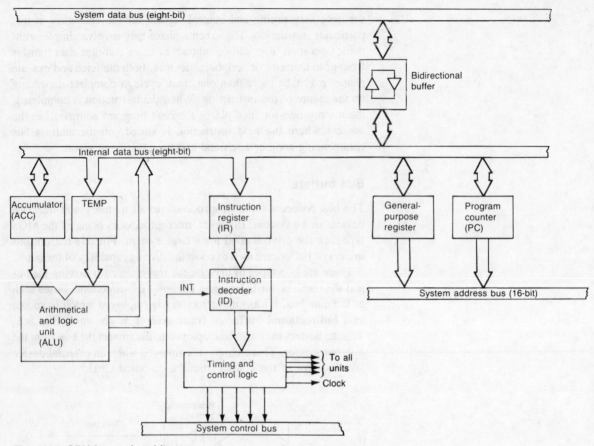

Fig. 9.6 CPU internal architecture

The only control signal that feeds into the CPU is the interrupt (INT) which goes into the instruction decoder to halt the operation of the microprocessor.

The arithmetic and logic unit (ALU)

The ALU has to carry out arithmetic and logic functions such as adding two numbers or performing a logic function such as NAND or NOR on two numbers. The ALU must therefore have two inputs: input A for the first number and input B for the second. These two numbers are first stored in two eight-bit registers: the **accumulator** (ACC) for input A and a temporary register, TEMP, for input B. When the ALU operation is completed, the result is stored in the accumulator replacing its original contents.

The instruction register (IR)

This is another eight-bit register which stores the coded instruction when it is fetched by the CPU. The instruction remains in the instruction register until it is executed before the next instruction is fetched and stored, and so on.

The instruction decoder (ID)

The coded instruction is an eight-bit word known as an operational code (op code). Each instruction such as add or store has its own unique code. Furthermore, each make of microprocessor chip has its own individual operational codes which are listed in an instruction set for that make of microprocessor. The stored coded instruction is fed into the instruction decoder from the instruction register. The decoder breaks down the code and instructs the timing and control logic to generate the necessary timing and control signals to execute the instruction.

The program counter (PC)

The microprocessor carries out its task in a predetermined sequence of smaller tasks known as the program. The program consists of a series of instructions. Each instruction consists of an operator (or opcode) and its associated data known as the operand. The instructions are arranged in a logical sequence and stored in successive memory locations known as program addresses, as shown in Fig. 9.7. To keep track of the program and ensure the CPU receives the instructions in the order intended by the program, a program counter, PC is used. The PC is a 16-bit register which holds the address of the next program location. For example, consider the program in Fig. 9.7. To commence the execution of the program, the CPU places the starting address, 0A20 (where the OPERATOR 1 is stored) in the program counter which in turn places it onto the address bus to fetch the first instruction. The first instruction is fetched into the CPU via the data bus and held in the instruction register. When this happens, the program counter is incremented to 0A20+1 = 0A21. OPERATOR 1 has no associated data and may then be executed immediately. When the first instruction is completed, the CPU causes the program counter to place its contents (0A21) onto the address bus to fetch the second instruction and store it in the instruction register. When that happens, the program counter is incremented to 0A22 (0A21+1) containing the associated data (OPERAND 2). OPERATOR 2 is decoded but cannot be completed without its associated data, which may be a number to be stored in a location or added to another number already in the accumulator. OPERAND 2 is fetched next by placing the contents of the program counter, 0A22 onto the address bus. OPERAND 2 now appears on the data bus upon which the program counter is incremented to 0A23 (0A22+1). When the second instruction is completed, the program counter places its new contents, 0A23 on the address to fetch the third instruction, and so on. Note here that an instruction may contain more than one 8-bit operand, as shown

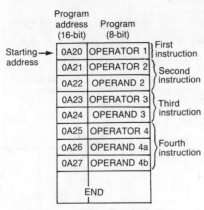

Fig. 9.7 Program construction

in the fourth instruction in Fig. 9.7. Two 8-bit operands are necessary, for instance, to accommodate the two bytes of a 16-bit address.

General-purpose registers

Every microprocessor has a number of general-purpose registers, (mainly 8-bit although some may be 16-bit width) for temporary storage of data or addresses in the course of program execution.

The UART

The universal asynchronous receiver–transmitter, UART (sometimes also known as the asynchronous communication interface adaptor, **ACIA**), has two modes of operation:

(*a*) the **transmit mode** in which the UART takes parallel data and converts it into a serial bit stream with start, stop and parity characters; and

(*b*) the **receive mode** in which it takes a serial bit stream and converts it to parallel data.

Figure 9.8 illustrates the connections of the UART with the CPU on one hand and a peripheral device such as a printer. The CPU determines the mode of operation of the UART. In the transmit mode, the UART receives parallel data from the CPU and converts them into serial data for the peripheral device. In the receive mode the reverse operation takes place, namely serial data from the peripheral device are converted into parallel data for the CPU. To ensure that the transfer of data takes place when the peripheral device is ready to receive data, **handshaking** is employed. Before the CPU begins to send information to the peripheral device, it enquires if the device is ready to receive data, by sending a **request to send** message. When the device is ready, it sends back the message **clear to send**. The CPU then proceeds to

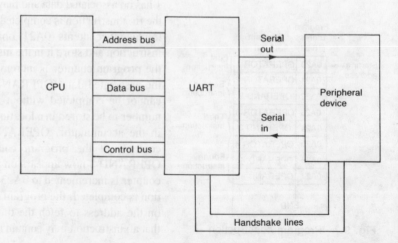

Fig. 9.8 UART connection

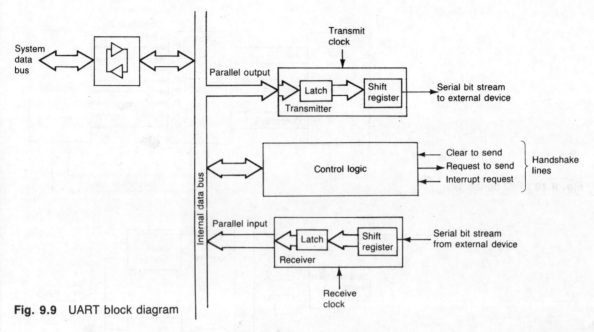

Fig. 9.9 UART block diagram

unload data into the UART registers. Conversely, when the peripheral device wishes to send data to the CPU, an interrupt request, IRQ, message is sent to the UART and subsequently to the CPU.

The basic functional block diagram consists of three main elements: a receiver, a transmitter and a control logic as shown in Fig. 9.9. The transmitter consists of a latch to hold the parallel data before conversion and a shift register to carry out the parallel to serial conversion. Similarly, the receiver consists of a shift register to carry out the serial to parallel conversion and a latch to hold the parallel data stable until the CPU requires the information. The UART is fully programmable with the control unit setting the chip into the transmit or the receive mode.

The UART as its name suggests, is an asynchronous device requiring a start and stop bit for each digital word transmitted to and from the CPU. Synchronous operation may be employed where start and stop bits are not required. A **USART**, universal synchronous/asynchronous receiver/transmitter provides a facility for synchronous operation.

The PIO

The parallel input/output (PIO) chip provides parallel lines known as ports to allow the CPU to interact with peripheral devices which require parallel data bits. Figure 9.10 shows the connections between the CPU, the PIO and a peripheral device such as a VDU. Only one eight-bit input/output port $P_0 - P_7$ is shown, although normally two or three eight-bit ports are provided. The I/O port is bidirectional in which each bit may be programmed individually to be an output

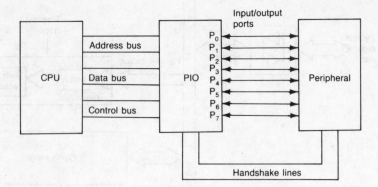

Fig. 9.10 PIO connection

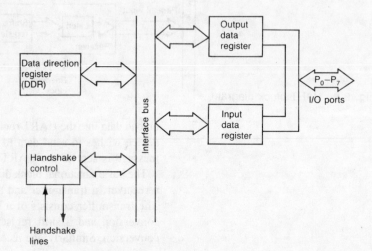

Fig. 9.11 PIO block diagram

or an input bit. Handshake lines are employed in the same way as those used in the UART.

Programming of the I/O port may be achieved by arranging the ports to look like a normal memory address. A part of the memory space is thus used exclusively for the operation of the I/O unit. This technique is known as memory mapped I/O. It is used by some microprocessors such as the 6502 and the 6800. Other microprocessors such as the Z80 and the 8080 employ a special I/O instruction to instruct the PIO to carry out the transfer of data between the microprocessor system and the peripheral device. Alternatively, the transfer of data may be carried out without the direct intervention of the CPU itself. This technique, known as **Direct Memory Access, DMA**, employs a **DMA controller** chip which provides a very fast data transfer.

A basic functional block diagram of the internal structure of a memory mapped PIO chip is shown in Fig. 9.11. The data direction register (DDR) is an eight-bit register which is used to define on a bit-by-bit basis the direction of each individual PIO port bit. Each bit in the data direction register specifies whether the corresponding bit in the PIO port is an input or an output. For example, in the 6502 microprocessor systems a logic 0 in the DDR specifies an I/O port bit an input while a logic 1 specifies an output. Thus to define all PIO bits as inputs, 00 (hex) is entered into the data direction register. Conversely, FF (hex) in the DDR will make all PIO bits into outputs. Further, OF (hex) will turn P_0-P_3 into inputs and P_4-P_7 into outputs, and so on. The output data register holds the parallel data stable to be read by the peripheral device, and the input data register holds the incoming parallel data stable till the CPU is ready to read it.

The instruction set

The program of a microprocessor system is a series of instructions which breaks down each operation into a number of individual tasks. These instructions are fed into the microprocessor chip in the form of an eight-bit binary number known as the machine code instruction or operational code (opcode) together with its associated data known as the operand. Each different make of microprocessor chip has its own set of opcodes known as the **instruction set**. Writing programs directly in machine code is a very lengthy and tedious process. Normally programs are written in a language which uses normal alphabetical letters and words. This is then translated into the appropriate series of opcodes and operands. The simplest form of translation is the **assembler** which employs the assembly programming language. In the assembly language each opcode is given a mnemonic name such as LDA for **load accumulator**, ADC for **add with carry** and JMP for **jump**.

Instruction sets may be divided into three main categories or sub-sets.

1. **Data transfer** which involves the movement of data between the CPU and memory locations, for example:
(*a*) load accumulator with contents of memory (LDA);
(*b*) store contents of accumulator into memory (STA); and
(*c*) load register X with contents of memory (LDX).

2. **Arithmetic and logical** includes instructions to perform arithmetical and logical operations such as:
(*a*) add two numbers with carry (ADC);
(*b*) subtract two numbers with carry (SBC);
(*c*) logical AND on two numbers (AND);
(*d*) logical EX-OR on two numbers (XOR);
(*e*) logical shift right (LSR); and
(*f*) arithmetic shift left (ASL).

3. **Test and branch** provides the facility for the microprocessor to vary the sequence of program operation by jumping or branching to another part of the programme, for example:
(*a*) jump to subroutine (JSR);
(*b*) branch if result is minus (BMI);
(*c*) branch if equal (BEQ).

10 Peripheral devices and interface circuits

The purpose of any digital system, including microprocessor and computer systems, is to initiate or control a physical activity such as the speed of a motor, the operation of a relay or the temperature of an oven. Devices such as stepper motors, relays and LED displays are known as **output peripherals**. Conversely, a digital system invariably requires some input information necessary for its operation. Devices that provide such information such as keyboards, switches and a variety of transducers are known as **input peripherals**. Peripherals are therefore devices that allow the user to communicate with the digital system. Such devices, however, cannot be connected directly to these systems. Input or output interface circuits are necessary to link the peripheral to the input or the output of the digital system as appropriate (Fig. 10.1). System interface circuits such as voltage-level shifters may also be required when two different logic families are used together in one system. System interface circuits such as buffers and transceivers are also required when digital data has to be moved through wires from one unit to another if the length of the wires is in excess of 0.5 m for high-speed systems or a few metres for low-speed systems.

The most common interface circuits are the analogue-to-digital and the digital-to-analogue converters. The ADC is used when the input is in analogue form such as a voltage level representing temperature or pressure. The DAC is employed when the output peripheral is of the analogue type, such as a cathode ray tube.

This chapter will deal with a number of other peripheral devices and their interface requirements.

Fig. 10.1 Connection of peripheral devices to digital system

Mechanical switch

When a digital or microprocessor system is used for control, some of its input signals may be obtained by the operation of a mechanical switch or contacts such as push-button switches, keypads, keyboards or reed switches. Mechanical switches suffer from what is known as contact bounce caused by the spring action of the switch which results in several contacts being made before coming to rest (Fig. 10.2). In logic circuits this may be interpreted as several individual switch operations. Both the leading-edge and trailing-edge bounce last for a period between 10 and 20 ms. To prevent this from happening a debounce interface circuit is employed. The most common debounce interface circuit is shown in Fig. 10.3 using an S–R latch. S_1 is a

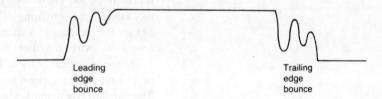

Leading
edge
bounce

Trailing
edge
bounce

Fig. 10.2   Switch bounce

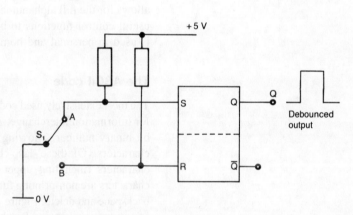

Fig. 10.3   Debounce circuit

Table 10.1   S–R truth table

S	R	Q
0	0	No change
0	1	0
1	0	1
1	1	Indeterminate

single-pole double-throw (SPDT) switch which when operated results in the contact moving from position A to position B and back again to A. The result is an ideal pulse at the output of the latch. As illustrated in Table 10.1, with the switch contact resting at A (before the switch is operated), S = 0 and R = 1, the latch is reset with an output of Q = 0. When the switch is operated, the contact moves to B and the latch is set (S = 1, R = 0), giving an output Q = 1. Contact bounce is obtained at both A and B. However, the S–R latch is set on the very first contact that the mechanical switch makes when it is moved to position B, producing a clean and fast transition from 0 V to 5 V at the output. The opposite is true when the switch shifts back to position B, resetting the latch and changing the output to 0 V.

Another solution to contact bounce is to wait for the status of the switch to remain stable for a period of say, 20 ms. This may be achieved by the use of a simple R−C low-pass filter followed by a Schmitt trigger.

Where a large number of switches are employed, e.g. in a keyboard, the hardware solution which requires a debouncer circuit for each individual switch becomes costly and cumbersome. In some cases a software-delay program is introduced which has the same effect as the R−C filter circuit.

The keyboard

A keyboard consists of two parts, a set of character keys which detect the pressure of a finger and close a switch and an encoder which converts the output of a pressed key into a unique binary code representing the character. The key switches are normally of the mechanical type consisting of a plunger which is moved by the finger against the pressure of a spring. At the end of its travel, the plunger forces two wires together, making contact.

There are several types of keyboards. The hex keypad consisting of 16 keys for the sixteen hex characters is used for small systems. The standard typewriter keyboard, known as the QWERTY keyboard allows for the full alphanumeric set of characters together with other useful control functions to be generated. The QWERTY keyboard is used in personal and home computers.

The ASCII code

The most extensively used code is ASCII, the American Standard Code for Information Interchange. ASCII represents a character by a seven-bit binary number, allowing for a maximum of 2^7 or 128 different characters. Of these 128 characters, 96 are the normal printing characters (including upper and lower cases). The remaining 32 characters are non-printing functions such as carriage return, line feed, backspace and delete. Figure 10.4 shows the full ASCII; it is arranged in columns and rows. To obtain the seven-bit code for a character read the most significant three bits in the column in which the character appears and the remaining least significant four bits in the row in which it appears. For instance, upper case R is in the 101 column and the 0010 row giving the ASCII code (MSB)101 and 0010(LSB) or 1010010. Similarly, carriage return (CR) has the code 0001101. ASCII is stored as eight bits by prefixing 0 to each code, thus upper case R will be stored as 01010010, and so on. This additional most significant bit can be used for parity checking.

Parity checking

Unlike analogue signals in which small errors may not significantly affect the operation of the system, a digital system may suffer a fatal fault as a result of the smallest possible error, namely a change of

b3	b2	b1	b0	b4=0 b5=0 b6=0	b4=0 b5=0 b6=1	b4=0 b5=1 b6=0	b4=0 b5=1 b6=1	b4=1 b5=0 b6=0	b4=1 b5=0 b6=1	b4=1 b5=1 b6=0	b4=1 b5=1 b6=1
0	0	0	0	NUL	DLC	SP	0	@	P	'	P
0	0	0	1	SOH	DC1	!	1	A	Q	a	q
0	0	1	0	STX	DC2	"	2	B	R	b	r
0	0	1	1	ETX	DC3	#	3	C	S	c	s
0	1	0	0	EOT	DC4	$	4	D	T	d	t
0	1	0	1	ENQ	NAK	%	5	E	U	e	u
0	1	1	0	ACK	SYN	&	6	F	V	f	v
0	1	1	1	BEL	ETB	'	7	G	W	g	w
1	0	0	0	BS	CAN	(	8	H	X	h	x
1	0	0	1	HT	EM	)	9	I	Y	i	y
1	0	1	0	LT	SUB	*	:	J	Z	j	z
1	0	1	1	VT	ESC	+	;	K	[	k	{
1	1	0	0	FF	FS	,	<	L	/	l	l
1	1	0	1	CR	GS	-	=	M	]	m	}
1	1	1	0	SO	RS	.	>	N	Λ	n	
1	1	1	1	SI	VS	/	?	O	—	o	DEL

Control characters   Printing characters

Fig. 10.4   ASCII

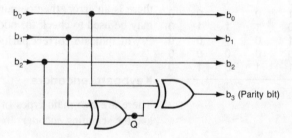

Fig. 10.5   Even parity generator

Table 10.2

b0	b1	b2	Q	b3 (parity bit)
0	0	0	0	0
0	0	1	1	1
0	1	0	1	1
0	1	1	0	0
1	0	0	0	1
1	0	1	1	0
1	1	0	1	0
1	1	1	0	1

the logic level of a single bit. Such errors may occur when data are transmitted from one unit to another. For this reason parity checking is employed. Parity checking is a method of error detection that employs an extra bit known as the parity bit.

There are two types of parity checking: even and odd. Before transmission, each word is examined by a parity generator or encoder which calculates the number of 1s in the word. If the number of 1s is odd, a parity bit at logic 1 is generated to make the number even for **even parity**. Conversely, a parity bit of 0 is generated to make the number odd for **odd parity**. Following transmission, a parity checking circuit is used at the receiving end to examine the received word to see if the correct parity still exists. This method can only detect the presence of an error in a word. It cannot detect the faulty bit. To do this, more complicated parity techniques have to be employed such as double parity checking or cyclic redundancy coding.

An even parity generator circuit is shown in Fig. 10.5. As can be seen from the truth table (Table 10.2), the logic level of the parity

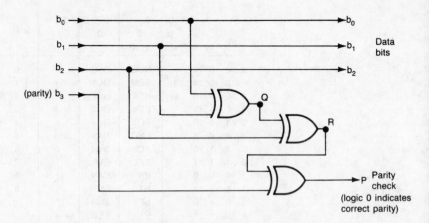

Fig. 10.6 Parity decoder

Table 10.3

b_0	b_1	b_2	b_3	Q	R	P
0	0	0	0	0	0	0
0	0	1	1	1	1	0
0	1	0	1	1	1	0
0	1	1	0	1	0	0
1	0	0	1	1	1	0
1	0	1	0	1	0	0
1	1	0	0	0	0	0
1	1	1	1	0	1	0

is such that the number of 1s in the four-bit output, b_0, b_1, b_2 and parity bit b_3 is always even.

A logic circuit for a parity checker or decoder is shown in Fig. 10.6. As can be seen from the truth table (Table 10.3), a logic 0 at the parity check output P indicates correct even parity. If any of the bits suffers a change in its logic level following transmission, i.e. if there is an error, then output P will indicate a logic 1. The decoder may be used to check for odd parity. In this case a logic 1 at output P will indicate correct parity.

Keyboard encoders

There are two main types of keyboard encoders: the static encoder and the scanning encoder. In the **static encoder** (Fig. 10.7) each key

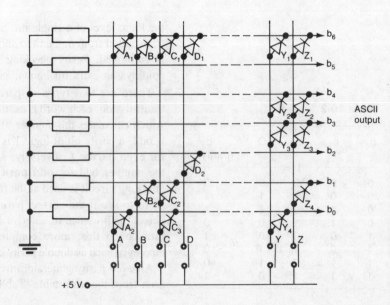

Fig. 10.7 Static keyboard encoder

has a separate connection and when pressed generates the appropriate code directly via the encoder. The encoder is a ROM array with the pattern of the diodes determining the seven-bit output code, b_0-b_6. With no key pressed, the output bits are at logic 0. If push-button keyswitch A is activated, diodes A_1 and A_2 are forward biased taking b_6 and b_0 to logic 1. The other bits (b_1-b_5) suffer no change in their logic level (i.e. they remain low) thus giving an output of 100001 (ASCII code for upper case A). Similarly for keyswitch Y. In this case diodes Y_1, Y_2, Y_3 and Y_4 conduct, forcing b_6, b_4, b_3 and b_0 to logic 1, giving an output of 1011001, and so on. In theory any number of keys can be used. However the number is limited by the number of connections that can be accommodated on the ROM decoder. For instance, to accommodate the 96 printing characters, 96 separate connections are necessary.

Scanning encoder

In the scanning encoder, the keyswitches are arranged in a matrix format. Using this arrangement the number of necessary connections are markedly reduced. For instance, an 8×8 matrix (Fig. 10.8) can provide 64 keyswitches with only $8+8 = 16$ connections. The switch matrix is connected to a keyboard encoder as shown in Fig 10.9. The encoder, known as a scanning encoder, is a read-only memory chip which generates the appropriate ASCII when a key is pressed in the switch matrix. The 8×8 switch matrix consists of eight vertical wires (x_0-x_7) and eight horizontal wires (y_0-y_7). At each of the 64 crosspoints is a push-button keyswitch which, when pressed, makes a connection between a vertical and a corresponding horizontal line.

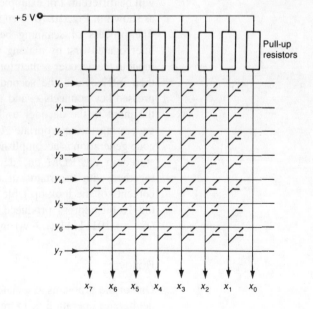

Fig. 10.8 8×8 keyswitch matrix

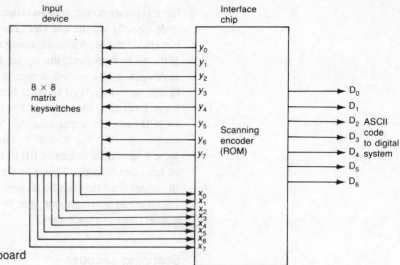

Fig. 10.9 Scanning-type keyboard encoder

In this way the particular key is identified by the encoder which then generates a unique address on its data output pins. The traditional method for key identification is the **row-scanning technique** (hence the name scanning decoder) in which each row is interrogated in turn to establish which key, if any, is pressed. Vertical lines, x_0-x_7, are taken to logic 1 via pull resistors and used as input data to the decoder (Fig. 10.8). The decoder scans the switch matrix by placing a logic 0 on each row in sequence starting with y_0 followed by y_1, and so on, in a circular manner. With no keyswitch pressed, the input to the decoder is 11111111 regardless of which row is being interrogated. However, when a key is operated the input to the decoder, x_0-x_7, will be different. For example if key K14 in the second row is closed as shown in Fig. 10.10, then row y_1 will be connected to column x_4. The process of scanning begins with the top row (at y_0). It is interrogated first by making $y_0 = 0$ (y_1-y_7 remain at logic 1). The input to the decoder is therefore unchanged at 11111111. The decoder then interrogates the second row by making $y_1 = 0$. Since the pressed key connects x_4 and y_1, line x_4 is forced to logic 0 changing the input to the decoder to 111101111. The decoder responds by generating the appropriate ASCII code on data pins D_0-D_6. This code generation is accomplished by the use of an ASCII look-up table stored in the ROM encoder. The key identification bits x_0-x_7, together with the status of rows y_0-y_7, are used to address the contents of the look-up table. In the example shown in Fig. 10.10 the 16-bit address produced by operating key K14 is 111101111 (x_0-x_7) 10111111 (y_0-y_7) in binary or F7BF in hex.

Rollover

One of the problems associated with keyboards is the accidental (or deliberate) operation of more than one key at a time. this is known

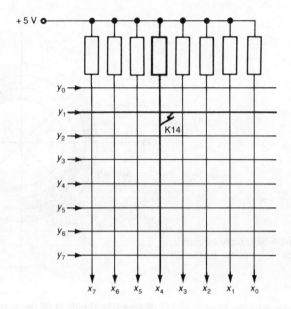

Fig. 10.10

as rollover. It is essential to detect a rollover condition to prevent wrong codes from being generated.

Absolute shaft encoder

A shaft encoder allows the angular position of a shaft to be determined by converting the angular movement (i.e. the position of the shaft) into a digital output. This is achieved by using a coded disc acting as a digital encoder attached to the shaft, as shown in Fig. 10.11. The coded disc is inscribed with a pattern consisting of a number of concentric tracks, one track for each bit of the digital output. The tracks are read individually by means of sensing heads. A three-bit binary encoding disc is illustrated in Fig. 10.12. There are three main types of encoder sensing heads: contact, magnetic and optical. In the **contact-type** encoder, actual contact is made between the encoded track and the sensor by means of a brush. This type of encoder suffers from all the problems associated with brushes such as friction, wear, dirt, arcing, contact resistance and vibration. The **magnetic-type** encoder employs a disc coated with a magnetic material on which a pattern has been pre-recorded in the form of magnetic spots. These spots are then read by pick-up heads which sense the presence or otherwise of a magnetic field. The **optical encoder** is the most common type of shaft encoder owing to its high accuracy and long lifespan. The optical encoder consists of a disc with a number of concentric tracks with each of the tracks being divided into a number of dark and transparent sectors. On one side of the disc are a number of LEDs (or some other light sources), one for each track. On the other side of the disc are an equal number of photoelectric sensors such as photo-transistors, one for each light source as shown in Fig. 10.13. Thus, for any position of the shaft and therefore the disc, a

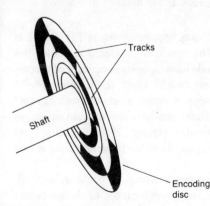

Fig. 10.11   Shaft encoder

MATTHEW BOULTON
COLLEGE LIBRARY

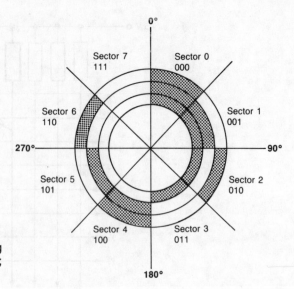

Fig. 10.12 Three-bit binary encoding disc. Outer track: 2^0; middle track: 2^1; inner track: 2^2

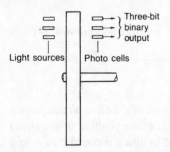

Fig. 10.13 Optical shaft encoder

Table 10.4

Sector	Angle	Binary code
0	0°–45°	000
1	45°–90°	001
2	90°–135°	010
3	135°–180°	011
4	180°–225°	100
5	225°–270°	101
6	270°–315°	110
7	315°–360°	111
0	0°–45°	000 and so on

unique combination of the sensors detects the light beam depending on whether or not there is a transparent sector between each light source and its corresponding photoelectric detector.

The resolution of the shaft encoder is determined by the number of tracks on the disc. With three tracks a three-bit output is obtained, giving a resolution of

$$\frac{360°}{2^3} = \frac{360°}{8} = 45°.$$

If four tracks are used, a four-bit output is obtained giving a resolution of

$$\frac{360°}{2^4} = \frac{360°}{16} = 22.5°.$$

In general

Resolution $= \dfrac{360°}{2^n}$, where n is the number of tracks.

The natural or direct binary coding system shown in Fig. 10.12 is not suitable for coded discs. The reason is illustrated in Table 10.4 which shows a natural binary code for a three-bit encoder. Each sector represents an angular movement of 45° which is the resolution of a three-bit encoder. As can be seen, movement between sectors 1 and 2 causes two bits to change state; similarly for movement across sectors 5 and 6. On the other hand, transition between sectors 3 and 4 causes all three bits to change; similarly for transition from sector 7 to sector 0.

In practice, since the sensing heads cannot be perfectly aligned, when two or more bits change state, one bit tends to change before

Table 10.5   Gray code

Sector	Angle	Gray code
0	0°–45°	000
1	45°–90°	001
2	90°–135°	011
3	135°–180°	010
4	180°–225°	110
5	225°–270°	111
6	270°–315°	101
7	315°–360°	100
0	0°–45°	000 and so on

Table 10.6

Decimal	Binary code	Gray code
0	0000	0000
1	0001	0001
2	0010	0011
3	0011	0010
4	0100	0110
5	0101	0111
6	0110	0101
7	0111	0100
8	1000	1100
9	1001	1101
10	1010	1111
11	1011	1110
12	1100	1010
13	1101	1011
14	1110	1001
15	1111	1000

the other. For instance, the transition from 001 (sector 1) to 010 (sector 2) may be produced as the sequence 001, 000, 010. A spurious code 000 is thus generated which, although it is present for a very short time, can in some applications be very troublesome. To avoid these problems, a different code, the **Gray code** is used. A Gray coded disc has the property that only one bit changes state as the shaft moves from one sector to the other. A Gray encoded three-track disc is shown in Fig. 10.14 with its Gray codes shown in Table 10.5. Conversion between the binary and the Gray codes for a four-bit system is shown in Table 10.6. Unlike the natural binary code, the Gray code does not have a specific weighting for each column and as such cannot be used for mathematical operations. For this reason whenever the Gray code is used an interface circuit is introduced between the shaft encoder and the digital system. This interface circuit converts the Gray code obtained from the detecting heads into a binary code for further processing by the digital system. A four-bit Gray-to-binary code converter circuit employing EX-OR gates is shown in Fig. 10.15. In some applications, binary-to-Gray code converters are required. A circuit that performs this function is shown in Fig. 10.16.

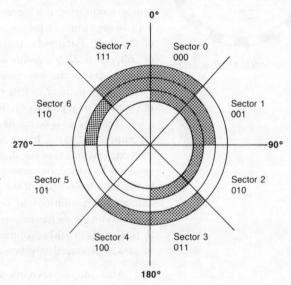

Fig. 10.14   Gray code encoder

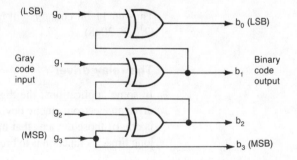

Fig. 10.15   Gray-to-binary code converter

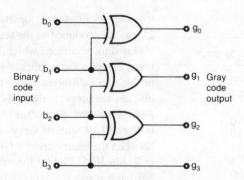

Fig. 10.16 Binary-to-Gray converter

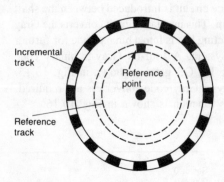

Fig. 10.17 Incremental shaft encoder disc

The incremental shaft encoder

Angular incremental encoders require a simpler disc than that used for absolute encoding. It consists of an incremental track equally divided into regular segments and a reference track with a single datum or reference segment as shown in Fig. 10.17. In the case of the optical type, light sources and photoelectric cells are used to detect the pattern on the disc. When the track is rotating, pulses are obtained from the sensor monitoring the incremental track. For an incremental track with 16 segments, 16 pulses are produced for one complete revolution of the shaft. Each pulse therefore represents an angular movement of $360°/16 = 22.5°$, giving an angular resolution of $22.5°$. A binary counter acts as the interface between the encoder and the digital system. The number of bits required at the output is determined by the number of segments on the incremental track. For 16 segments, a four-bit counter is adequate. A 64-segment track will require a six-bit output, and so on. In the case of a multi-turn encoder where more than one revolution of the shaft has to be recorded, a specific-position switch is provided to clear the counter before the start of the process. In this case the bit size of the counter is determined by the maximum number of revolutions to be recorded as well as the number of segments on the incremental track. Assuming a 16-segment track, then to record up to 100 revolutions of the shaft, the maximum number of pulses produced is given as

$$\text{Max. no. of revolutions} \times \text{no. of segments} = 100 \times 16$$
$$= 1600 \text{ pulses.}$$

Thus an 11-bit counter will be needed to represent the 1600 pulses ($2^{11} = 2048$).

The relay driver

In some applications, the digital or computer system is required to drive an electromagnetic device such as a solenoid or a relay. A relay is one of those devices that have been used in electronics for a very long time. Its function is to break, make or change one or more electric

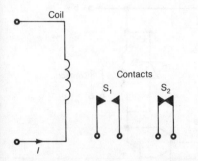

Fig. 10.18 Relay operating two contact switches: S_1 normally closed and S_2 normally open

contacts. Solenoids are extensively used for switching arrangements such as those used in telephone systems and automatic test equipment. A relay consists of a coil which when magnetised by a d.c. current creates a magnetic field which makes or breaks one or more mechanical contacts. Figure 10.18 shows a relay coil operating two separate contacts: S_1 (normally open) and S_2 (normally closed). When the coil is energised by a d.c. current, S_1 makes (closes) and S_2 breaks (opens). When the current ceases, the two contacts return to their normal states. The current required to operate a relay is far in excess of the standard 10 mA current supplied by an output buffer stage of the digital system. An interface circuit is therefore necessary to provide the driving current to operate the relay as shown in Fig.

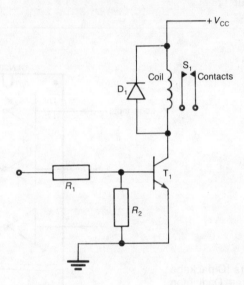

Fig. 10.19 Relay driver circuit

10.19. When a logic 1 is applied to the interface circuit, transistor T_1 becomes forward biased, drives a current through the relay coil and operates the relay. Contact S_1 closes. When a logic 0 is applied to the circuit, T_1 is turned off and its current drops to zero. The coil is de-energised and the relay returns to its normal state with contact S_1 open. D_1 is a protection diode inserted to protect the transistor from any overshoot voltage that may appear on its collector due to the back e.m.f. generated by the coil when the transistor turns off. The overshoot voltage may be of such magnitude as to exceed the maximum rateable collector voltage, thus resulting in damage to the transistor. The diode, which is normally reverse biased will conduct only if the collector voltage exceeds that of the supply voltage V_{CC}. If that happens, D_1 conducts as shown in Fig. 10.20. Diode current I_D ensures that the back e.m.f. is depleted rapidly.

A single transistor may not provide adequate current to drive the relay which may need as much as 250 mA. For this reason a Darlington pair is normally employed as shown in Fig. 10.21. T_2 and

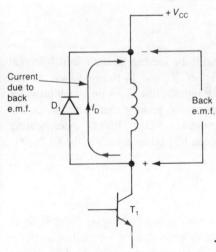

Fig. 10.20 Operation of protection diode D_1

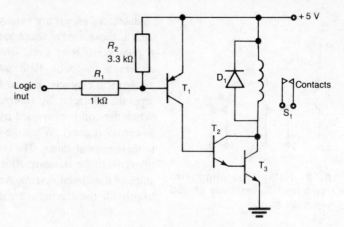

Fig. 10.21 Relay driver interface circuit with initial amplifer T_1

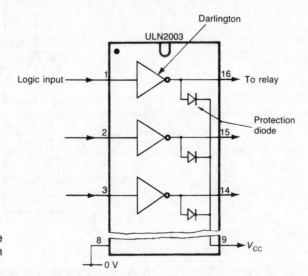

Fig. 10.22 Relay drivers IC package consisting of seven separate Darlington pairs with protection diodes

T_3 form a Darlington pair with T_1 serving as their initial driver. When the input is high, T_1 is off. T_2 and T_3 have no base current and thus remain off. When the input is low, T_1 is on which provides base current for the Darlington pair, turning them on. The coil is energised and the relay operates. Relay drivers incorporating protection diodes are available in IC packages such as the ULN2003 shown in Fig. 10.22.

Alphanumeric displays

Alphanumeric displays are commonly used to display the reading of a digital measuring instrument or other data of a digital or a micro-processor system. A number of techniques may be used to display numbers and/or some or all of the alphabets as well as any other

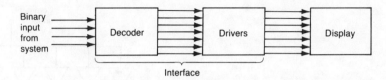

Fig. 10.23   A basic display system

characters such as question and exclamation marks. The most common type is the seven-segment display which is suitable for denary as well as hexadecimal characters. Another technique is the dot matrix which can provide enhanced quality and increased range of characters. A basic display system is shown in Fig. 10.23. The binary output of the digital system must first be decoded into a digital form suitable for the type of display used and then fed into a driver stage before going into the display device. The interface circuit therefore consists of two basic components: a decoder and a driver.

Seven-segment display

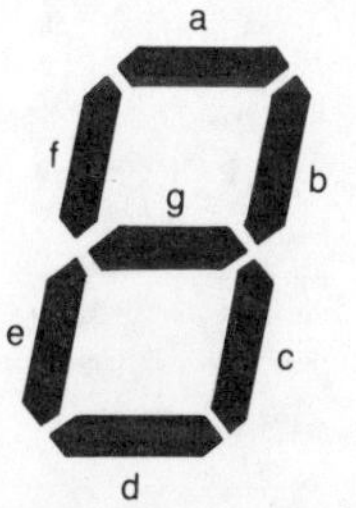

This type of display consists of seven separate segments labelled **a** to **g** as shown in Fig. 10.24. Seven segments are the minimum number of segments that may be used to represent the numbers from zero to nine as illustrated in Fig. 10.25. A number of the alphabet characters may also be represented by the seven-segment display. Of particular interest are the hexadecimal characters which may be displayed as illustrated in Fig. 10.26. There are two types of seven-segment displays: the **light-emitting diode** (LED) and the **liquid-crystal display** (LCD). The latter type requires very little power to operate, hence its use in portable equipment where power consumption is an important consideration. On the other hand, LED displays can be observed in the dark while the LCD requires ambient light.

Fig. 10.24   Seven-segment display

Fig. 10.25   Seven-segment representation of denary digits

0123456789

Fig. 10.26   Seven-segment representation of hex characters

AbCdEF

LED interface requirement

Figure 10.27 shows the interface for one seven-segment LED display. The LED converts electrical current into light. Thus, to illuminate one segment of the display, current must be directed to the diode making up that segment. As shown, one terminal of each diode is connected to a common point. In the example shown, the anodes are connected together making what is known as a **common-anode configuration**. When the cathodes are connected together instead of

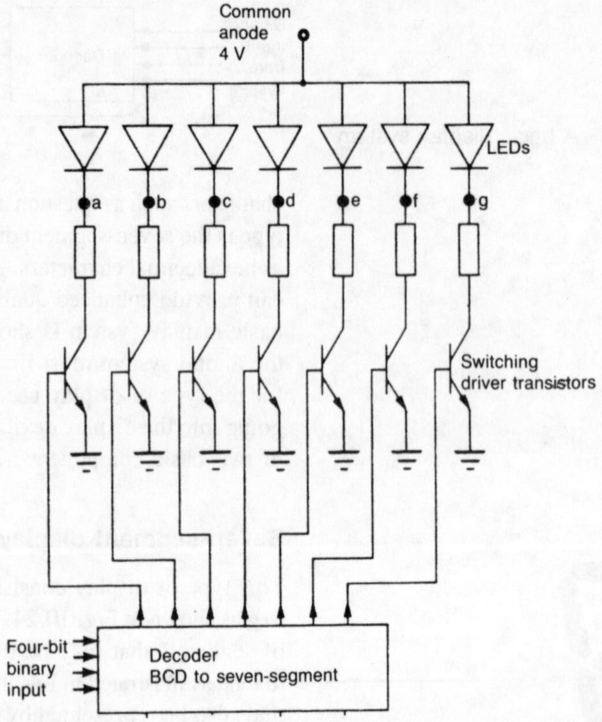

Fig. 10.27 LED display driving requirement

the anodes, the result is a **common-cathode configuration**. Each seven-segment display forms one digit of a complete multi-digit display. Each digit thus has eight terminals: one for each segment and one for the common connection. In some instances, a decimal point is included in the display, giving a ninth terminal.

Before the segments are driven, the output of the digital system must be changed into the appropriate signal to drive the display. The input from the digital system is in the form of a binary code (binary-coded decimal is commonly used) which is converted into a seven-line signal to drive each of the seven segments. This conversion is carried out by the decoder. For instance, if the character 1 is to be displayed, then output lines b and c from the decoder will be taken to logic 1 to switch their respective driver transistors on to illuminate segments b and c of the display. The other output lines of the decoder remain at logic 0, preventing the other segments from conducting and so on for the other characters as illustrated in Fig. 10.28. Where a decimal point (DP) is required, an eight-segment display is employed together with a binary-to-eight-line decoder.

In practice, the switching driver transistors are contained in a single IC package. Furthermore, the functions of the decoder and the driver may be performed by a single IC, e.g. the 7447 for common-anode and the 7448 for common-cathode displays. One such display is shown in Fig. 10.29.

Input count	Segments activated							Resulting display
	a	b	c	d	e	f	g	0
0	a	b	c	d	e	f		0
1		b	c					1
2	a	b		d	e		g	2
3	a	b	c	d			g	3
4		b	c			f	g	4
5	a		c	d		f	g	5
6			c	d	e	f	g	6
7				d	e	f		7
8	a	b	c	d	e	f	g	8
9	a	b	c			f	g	9
A	a	b	c	d	e	f		A
B			c	d	e	f	g	b
C	a			d	e	f		C
D		b	c	d	e		g	d
E	a			d	e	f	g	E
F	a				e	f	g	F

Fig. 10.28

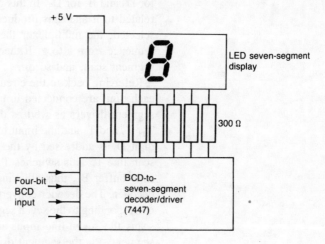

Fig. 10.29 The use of decoder/driver package

Multiplexing

So far we have considered a single digit and its interface requirement. In multi-digit displays, a separate interface circuit has to be constructed for each digit. A large number of connections therefore become necessary making the circuit complex and expensive. For instance, a six-digit display will need $6 \times 8 = 48$ connections. Moreover, a separate driver is required for each segment which adds to the complexity of the circuit and its cost. To overcome these problems, use is made of multiplexing or time-sharing techniques. Instead of continuously driving each individual digit, the digits are multiplexed, i.e. energised one at a time and in sequence; the first digit is followed by the second and so on to the last digit. The process is then repeated starting at the first digit. Provided the frequency of the multiplex is high enough, the brightness of the LEDs will be perceived as continuous by the human eye even though only having an illuminating current for part of the time. Another advantage of multiplexing is the reduced power consumption that results from employing a pulse of a short duration to drive the LED. The effect of multiplexing on brightness is very small. This is because of the peculiarities of the human eye which perceives pulsed operations as low at 10% of the time as having the same brightness as continuous and uninterrupted illumination. Multiplexing may also be applied to the segments as well as to the digits in which the segments are addressed one at a time and in sequence. Segment multiplexing further reduces the power consumption of the display and improves its efficiency.

A fully multiplexed LED digital display interface circuit is shown in Fig. 10.30. Only three digits are shown for simplicity. First the multiplexer unit scans the digits in sequence: D_1, D_2, D_3 and then back to D_1, and so on. With each selected digit, the multiplexer enables the input buffer associated with that digit, i.e. B_1 for D_1, B_2 for D_2 and B_3 for D_3. In this way, each digit receives the information related to it and thus produces the correct display. After a digit is selected, the multiplexer then scans the segments of that digit in sequence from a to g. It then selects the next digit and repeats the segment scan, and so on.

Referring back to the circuit in Fig. 10.30, identical segments of each digit are connected in parallel. The multiplexer addresses the segment drivers as well as the digits through switching transistors T_1, T_2 and T_3 and the input tri-state buffers B_1, B_2 and B_3. The first digit D_1 is addressed by the multiplexer placing a logic 1 on digit scan line 1. This switches T_1 on and at the same time enables tri-state buffer B_1, thus allowing D_1 binary data to go through to the decoder. The decoder converts the binary data to a seven-line signal corresponding to the seven segments of the digit. The multiplexer then scans the seven-line input and feeds each one to the appropriate segments via the segment drivers in sequence a to g. When all the segments are addressed T_1 is turned off and T_2 is turned on to address the second digit by placing a logic 1 on digit scan line 2. This enables tri-state buffer B_2 allowing digit D_2 binary data through to the decoder and consequently the seven-line output to the multiplexer. The multiplexer then scans the segments' signals and feeds them to the appropriate segment of digit D_2 via the segment drivers in sequence a to g. At the end of the scan, T_2 is turned off and T_3 is turned on and so on until the last digit is addressed. The process is then repeated, starting with the first digit. The speed of multiplexing is determined by the clock whose frequency is chosen taking into account the number of digits in the display.

LCD interface requirement

The interface for a liquid crystal display differs from the LED type in the drive requirement for the segments. Liquid crystal displays

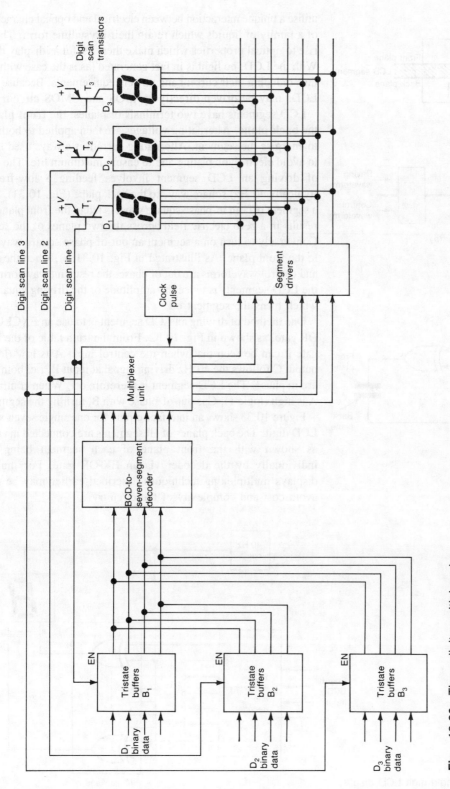

Fig. 10.30 Three-digit multiplexed display

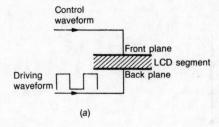

(a)

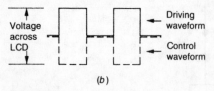

(b)

Fig. 10.31

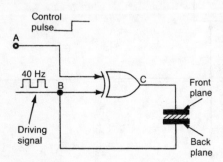

Fig. 10.32

utilise a unique interaction between electrical and optical characteristics of a family of liquids which retain their crystalline form. They give rise to optical properties which make them useful as display devices. With the LCD, no light is in fact generated (as is the case with LEDs) thus reducing their current and power requirements. Because of this, LCDs may be driven directly by MOS and CMOS circuitry.

LCD segments have two terminals or planes: the **front plane** and the **back plane**. Alternating voltages are then applied to both planes to energise the segment. Alternating current is always used in order to avoid electrolytic plating and to ensure maximum life. The method of driving an LCD segment involves feeding a low-frequency (typically 40 Hz) square wave to the back plane (Fig. 10.31). To turn a segment off, an in-phase square wave is fed to the front plane which results in a zero electric field across the two planes of the segment. Conversely to turn on a segment an out-of-phase square wave is fed to the front plane. As illustrated in Fig. 10.31(b), when the driving and control waveforms are out of phase, the resultant waveform across the LCD segments is twice the amplitude of the driving square wave which turns the segment on.

One method of driving an LCD segment is to use an EXCLUSIVE-OR gate, as shown in Fig. 10.32. From the truth table of the EX-OR gate it can be seen that when the control input A is low (logic 0), output C follows the 40 Hz driving signal at input B, i.e. both signals are in phase. The LCD segment is therefore off. When control input A is high (logic 1), C is out-of-phase with B, turning the segment on.

Figure 10.33 shows an interface circuit for one single seven-segment LCD digit. The back planes of all segments are connected in common as shown with the front plane of each segment being driven individually by the decoder via an EX-OR gate. For multi-digit displays multiplexing techniques described earlier may be used to avoid cost and complexity of the circuitry.

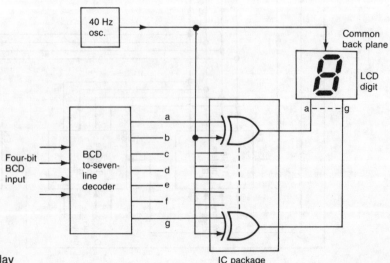

Fig. 10.33 Single-digit LCD display

Dot matrix display

A dot matrix display consists of a number of LEDs arranged in an array of rows and columns. The most common arrangement is the five-by-seven matrix shown in Fig. 10.34 comprising five columns and seven rows. This X-by-Y arrangement allows for X−Y addressing of the LED elements which can be used in conjunction with a ROM character generator. The five-by-seven matrix can be used to represent the complete alphanumeric characters, some of which are shown in Fig. 10.35.

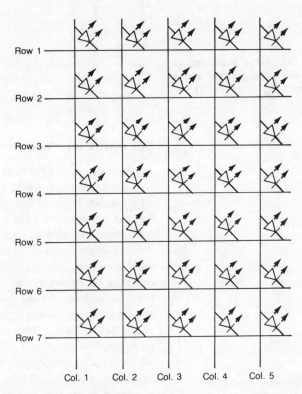

Fig. 10.34 Five-by-seven dot matrix display

The process of generating characters involves scanning of the rows (or the columns), selecting one at a time and energising the appropriate LED in that row (or column). The process is then repeated for the next row (or column), and so on. When all the rows (or columns) have been selected in the proper order, the procedure is repeated starting from the top row (or the first column). If the scanning frequency is fast enough (about 100 Hz), flicker-free characters will be formed. If the matrix is scanned from left to right, column by column, it is called horizontal scanning. When the scanning is carried out row by row, it is known as vertical scanning (Fig. 10.36). Horizontal scanning is suitable for displays of up to four characters, while vertical scanning is normally used for applications requiring five or more characters.

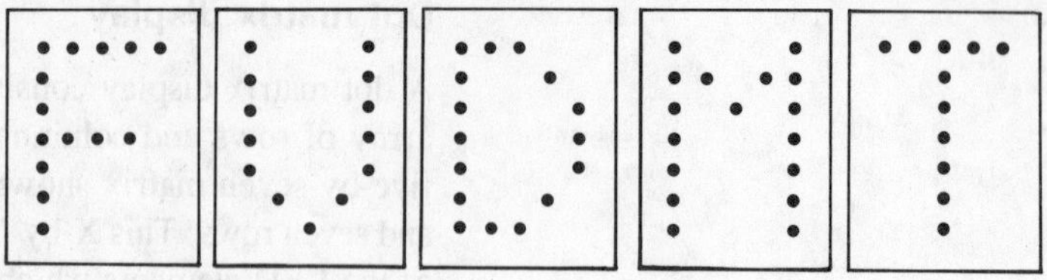

Fig. 10.35 Dot matrix character representation

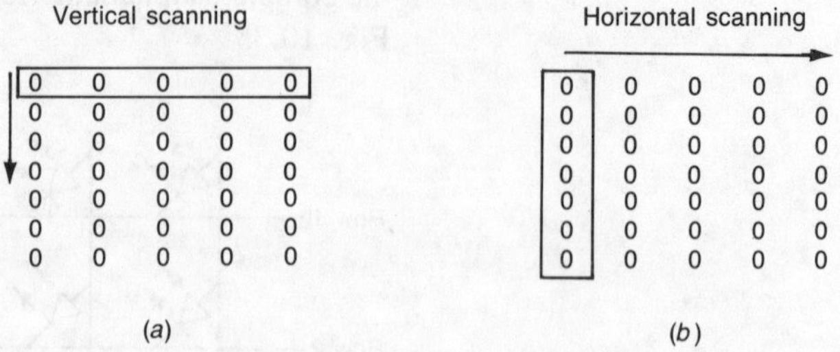

Vertical scanning

Horizontal scanning

Fig. 10.36 Dot matrix scanning: (*a*) vertical scanning (top to bottom), and (*b*) horizontal scanning (left to right)

(*a*)

(*b*)

Dot matrix interface circuitry

An interface circuit for a single-character dot matrix display employing horizontal scanning is shown in Fig. 10.37. This consists of four parts: an input data latch (or storage buffer), a ROM character generator, line and column drivers and a clock and timing control unit. The input latch stores the coded six-bit input data from a keyboard or other digital devices which could be of any binary code, although ASCII is the most common. A six-bit code is sufficient to cover all the characters that may be displayed with this type of array. The stored data are used to address the ROM character generator which provides sequential five-by-seven array information appropriate to the ASCII coded data. The line and column drivers provide the appropriate current to illuminate the LEDs, and finally the clock and timing unit which initiates and synchronises the operation of the circuit. The operation of the circuit is as follows. The six-bit ASCII input data is entered into the six-bit input latch and remains there until the data is updated. The contents of the latch are then used to address the character generator to produce a seven-line output which corresponds to the status of the LEDs in column 1. This seven-line signal is fed to the matrix via the row drivers. At the same time as the seven-line drive signal appears at the LEDs at column 1, the clock and timing unit triggers the column driver block and connects the first column to 0 V to complete the circuit. The other columns remain disconnected. In this way, the appropriate LEDs in column 1 will be lit. For instance if the character to be generated is A, then the logic states of the seven-line drive signal will be 0011111 starting at row 1. With column 1 connected to 0, the lower five LEDs of the first column will be illuminated. The ROM is then triggered by a timing signal to present the second column of character information at its seven-line output

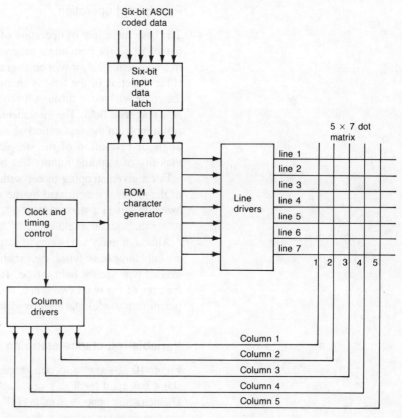

Fig. 10.37 A single-character dot matrix interface

while at the same time the column driver block is triggered to connect column 2 to 0 V. The appropriate LEDs in the second column are thus lit. The process is repeated for the remaining columns 3, 4 and 5. When this is completed, the whole cycle is repeated starting with column 1.

The stepping motor

The principle of operation of the stepping motor is the same as that of an ordinary a.c. motor, namely the creation of a rotating magnetic field by a current through a stator which causes a rotor to move. However, while the a.c. motor requires sinusoidal current to produce a continuous movement, the stepping motor requires pulse waveforms, with each pulse moving the rotor through a specified angle. The main use of stepping motors is in position control applications such as robotics, X−Y plotters and disc drives. Its advantage over other position control techniques is that the precision with which the motor turns for any given number of pulses obviates the need for feedback and eliminates much of the hardware requirement of a conventional closed-loop system.

Principle of operation

The basic principle of operation of a stepping motor is that when a bar of iron or a permanent magnet known as a rotor is suspended in a magnetic field, it will be aligned with the direction of that field. If the direction of the field is changed by the application of a pulse, the rotor will move through a precise angle to align itself with the new magnetic field. The incremental angular movement of the rotor resulting from the application of a single pulse is known as the **step angle** or **resolution** of the stepping motor. The step angle of the majority of stepping motors lies between $0.45°$ and $90°$.

For a given stepping motor with a given step angle, the position of the motor is determined by the number of pulses fed into it. For example, given a step angle of $6°$, after 15 pulses the motor would rotate through an angle of $15 \times 6° = 90°$.

Although many variations of stepping motors are available, there are only three basic types: the variable-reluctance type, the permanent-magnet type and the hybrid type. The hybrid stepping motor contains features of the other two types, resulting in high angular resolution (small step angle) and improved torque capabilities.

Variable-reluctance stepping motor

Figure 10.38 shows a 15-step variable-reluctance stepping motor. The stator has eight teeth or poles, each one with its own winding coil. Diametrically opposite windings (P_{1a} and P_{1b}, P_{2a} and P_{2b}, and so on) are connected in series so that when one tooth acts as a north pole, the other acts as a south pole. There are thus four independent pairs of stator poles known as phases: phase 1 (P_{1a}, P_{1b}), phase 2 (P_{2a}, P_{2b}), phase 3 (P_{3a}, P_{3b}) and phase 4 (P_{4a}, P_{4b}). The rotor has a different number of poles or teeth from the stator. Figure 10.38 shows three pairs of poles, a total of six poles, namely $R_{1a}-R_{1b}$, $R_{2a}-R_{2b}$ and $R_{3a}-R_{3b}$. As can be seen from the diagram, only one pair of the rotor poles can be in line with a pair of stator poles at any one time. The stator poles are placed at $45°$ ($360°/8$) while the rotor poles have $60°$ ($360°/6$) between them. This gives a step angle of

$$60° - 45° = 15°.$$

When phase 1 is energised with a direct current, a magnetic field along stator poles P_{1a} and P_{1b} is created. The rotor is therefore attracted into the position shown in Fig. 10.38 in which a pair of diametrically opposite rotor poles, R_{1a} and R_{1b} are aligned with the magnetic field. If phase 1 is switched off and phase 2 is switched on, the rotor will move $15°$ clockwise so that rotor poles R_{2a} and R_{2b} are aligned with the newly energised stator poles. A further $15°$ step movement will be obtained by switching phase 2 off and phase 3 on. By repetitively switching on the stator phases in the sequence 1, 2, 3, 4, 1, etc., the motor will rotate in the clockwise direction in

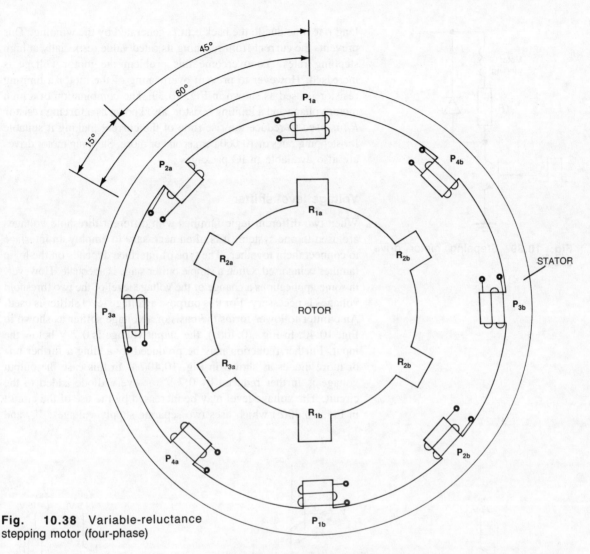

Fig.   10.38 Variable-reluctance stepping motor (four-phase)

incremental steps of 15°. If the sequence is reversed, i.e. 1, 4, 3, 2, 1, etc., the rotation will be anti-clockwise. The step angle may be changed by employing a different number of stator and rotor teeth. For instance, a step angle of 30° is obtained with six stator poles (three phases) and four rotor poles. Conversely, a higher resolution (a smaller step angle) may be produced by employing a larger number of stator and rotor poles.

Stepping motor drive requirements

The performance of a stepping motor is highly dependent on its drive system. The simple relay drive circuits (Figs. 10.19 and 10.21) described earlier in this chapter may be used to drive each stator winding. In practice, the current through the winding suffers from

MATTHEW BOULTON
COLLEGE LIBRARY

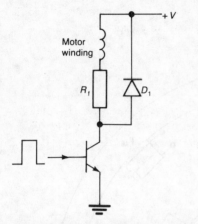

Fig. 10.39 Stepping motor drive circuit

long rise time due to the back e.m.f. generated by the winding. This prevents the current from reaching its rated value, especially at high stepping rates. To overcome this problem the motor voltage is increased. However to prevent overheating of the motor a limiting resistor is used as shown in Fig. 10.39. The combination of a high motor voltage and a limiting resistor, also known as a **forcing resistor** R_f, is used to reduce the rise time of the current making it suitable for stepping rates of 10 000 per second or more. Stepping motor drives are also available in IC packages.

Voltage level shifter

When two different logic families with different threshold voltages are used in one system, it is often necessary to employ an interface to connect them together. The type of interface depends on the logic families being used. Often a simple buffer stage is adequate. However, in some applications a change of the voltage level of the two threshold voltages is necessary. For this purpose a voltage level shifter is used. An emitter follower forms the basis of a voltage shifter as shown in Fig. 10.40. In Fig. 10.40(a), the output voltage is 0.7 V below the input. Further reduction may be produced by adding a further one or more diodes as shown in Fig. 10.40(b). In this case the output voltage is further reduced by 0.7 V for every diode added to the circuit. The voltage level may be increased by the use of the circuit in Fig. 10.40(c) which uses two separate supply voltages: V_{S1} and

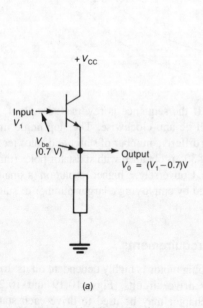

Fig. 10.40 Level-shifter circuits

V_{S2}. When the input voltage, $V_1 = 0$ V, T_1 is turned off. R_1 and R_2 function as a potential divider giving an output voltage,

$$V_0 = \frac{15}{10+1} = \frac{15}{11} = 1.36 \text{ V.}$$

For an input voltage of 2 V (TTL logic 1 threshold), T_1 conducts with its emitter at $2 - 0.7 = 1.3$ V giving an output voltage $C_0 = 15 - 1.3 = 13.7$ V. Negative logic may be obtained at the output by changing V_{S2} into a negative voltage.

The opto-isolator

The opto-isolator, also known as an opto-coupler is a photoelectric switching device consisting of an LED emitting infra-red light and a photo-transistor as shown in Fig. 10.41. For increased current facility, a photo-Darlington may be used in place of the single photo-transistor. Both the LED and the photo-transistor are contained in a sealed unit with two input and two output connections. When a logic 1 is applied to the input terminals, the LED conducts emitting infra-red light which switches the transistor on to produce an output current. To provide adequate current to drive the LED, a transistor amplifying stage may be used.

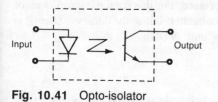

Fig. 10.41   Opto-isolator

11 CRT display

One of the most versatile display units is the cathode ray tube, CRT. The CRT may be used to display more characters and more lines than most other display devices. In addition, more dots per character may be utilised together with full graphic capabilities on specialised and more expensive units. In the microprocessor field, the cost of a visual display unit is of critical importance. For this reason microprocessor systems, especially the home computer, utilise the domestic television receiver as the visual display unit, VDU.

The CRT

The CRT operates on the same principle as the old thermionic valve, namely a negatively charged hot cathode emits electrons which are attracted to and collected by a positively charged anode.

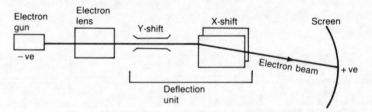

Fig. 11.1 The cathode ray tube (CRT)

In the CRT, high-speed electrons are emitted by an electron gun (Fig. 11.1). They are focused and accelerated by an electron lens and directed towards the screen which acts as the positively charged anode. The screen is coated from the inside with a fluorescent powder or phosphor which gives a visible glow when hit by the high-speed electrons. The electron beam generated by the electron gun gives a stationary dot on the screen. In order to produce a display, the CRT must have the capacity to deflect the beam in both the horizontal (X) and vertical (Y) directions.

Two methods are used to deflect the beam in a CRT: the electrostatic and the electromagnetic. The **electrostatic** method uses two plates facing each other with an electric potential across them. The electrostatic field thus created deflects the electron beam passing through it. Two pairs of deflecting plates are used: the X-plates for

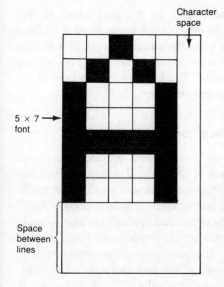

Character space

5 × 7 font

Space between lines

Fig. 11.2 6 × 10 character block on a CRT display

horizontal shift and the Y-plates for the vertical shift. By putting an electrostatic charge (voltage) across the plates, the electron beam may be made to deflect to any point on the screen. **Electromagnetic** deflection uses the magnetic field created by a current passing through a coil to deflect the beam. Two pairs of deflecting coils, known as scanning coils in TV receivers are placed at right angles to each other along the neck of the tube. Electromagnetic deflection is more suitable for high tube voltages and wide angles of deflection, hence its use in TV receivers.

The CRT alphanumeric display

The dot matrix format described in the previous chapter is used to display characters on a CRT screen. The most popular is the 5 × 7 in which each character is displayed as a 5 × 7 matrix, called a font within a block of 6 × 10 to provide for spaces between characters and lines as shown in Fig. 11.2. The VDU screen is then divided into columns and rows equivalent to character blocks and lines, respectively, as shown in Fig. 11.3. Typical VDU formats are 80 (columns) × 24 (rows), 40 × 24, 64 × 16 and 32 × 16.

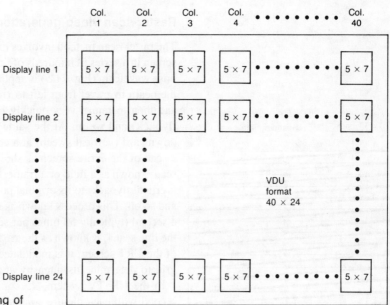

Fig. 11.3 VDU format comprising of 40 columns and 24 rows

The process of producing a display of characters on a CRT screen involves generating the appropriate dots of each character in the correct block on the screen. This process consists of two main parts: positioning (i.e. deflecting the beam to the correct position for a particular dot of a particular character) and modulating the beam (i.e. controlling the brightness of the beam). Modulating the beam is

MATTHEW BOULTON
COLLEGE LIBRARY

achieved by controlling the voltage on the grid which in turn determines the brightness of the dot. Turning the beam off is achieved simply by feeding a negative **blanking pulse** to the grid. Positioning the beam may be carried out in two different modes: the co-ordinate mode and the raster-scan mode. The latter method is used in TV receivers and is by far the more popular.

Co-ordinate mode of video generation

In the co-ordinate mode two binary inputs are used directly to determine the precise position of the beam. The two binary inputs are applied to two digital-to-analogue converters which transform the binary inputs into a voltage proportional to their respective digital values. These analogue voltages are then applied, via amplifiers, to the X and Y deflection plates (or deflection coils in the case of electromagnetic deflection). The beam is simultaneously turned on or off as necessary depending on the character or graphic representation. The dots are plotted rapidly to avoid flicker due to fading of the image on the screen. The whole display is refreshed at least 50 times a second to maintain a continuously visible display.

Raster-scan video generation

The raster-scan method involves driving the electron beam across the screen in a series of hundreds of lines as shown in Fig. 11.4. A **timebase** waveform (Fig. 11.5) is applied to the X plates (or coils) causing the beam to travel from left to right at a steady speed (**the sweep**) and then to return very quickly back to the left of the screen (**the flyback**) and so on. At the same time, the beam is made to travel downward (vertical sweep) at a constant but slower rate to produce a scan of the entire screen as shown in Fig. 11.4. At the end of the scan, known as a **field** or a frame, the beam is quickly made to return (vertical flyback) to its original position ready to scan the next field, and so on. This process known as **scanning** is repeated several times a second (typically 50 times per second). The path of the beam, i.e. the line scans, is known as the **raster**. As the beam scans the surface of the CRT screen, it is modulated by applying a video signal to the grid to determine its intensity at any particular point.

In the UK TV system, 25 complete pictures are scanned every second with each picture containing 625 lines. To avoid the flicker associated with such a slow picture repetition without increasing the bandwidth of the system, **interlaced scanning** is employed: the 'odd' lines are scanned first followed by the 'even' ones, and so on. Each half picture is known as a frame or a field, giving a field frequency of **50 Hz** and a line frequency of $625 \times 50 = 15\,625$ Hz or **15.625 kHz**. In some VDU applications, interlacing is dispensed with. In such cases 25 fields (complete pictures) are scanned a second to give a field frequency of 25 Hz with the line frequency remaining unchanged at 15.625 kHz.

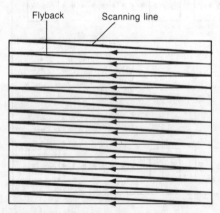

Flyback   Scanning line

Fig. 11.4 Scanning of a CRT screen

Sweep   Flyback

Fig. 11.5 Timebase waveform

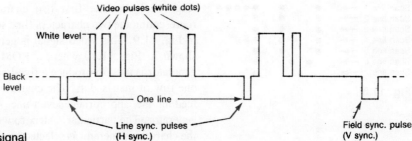

Fig. 11.6 Composite video signal

The complete video signal, known as composite video, consists of three components as shown in Fig. 11.6:

1. video information line by line;
2. line or horizontal synchronising pulses (H sync) to indicate the end of one line and start of the next; and
3. field or vertical synchronising pulses (V sync) to indicate the end of one field and start of the next.

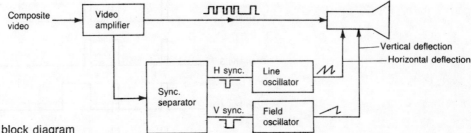

Fig. 11.7 VDU block diagram

The composite video is applied to a VDU to produce a display (Fig. 11.7). At the video amplifier stage, the sync. pulses (line and field) are separated away from the line video information and used to trigger the line and field timebase oscillators respectively to produce a raster scan on the screen of the CRT.

Display generating system

The purpose of a display generating system is to produce the appropriate composite video signal and apply it to the VDU. It acts as the interface between the microcomputer and the VDU. The display consists of a number of characters arranged in a number of rows. Assuming a 5 × 7 dot format, then each display line of characters will occupy seven scanning lines. Assuming a non-interlaced system, the process of scanning involves the electron beam sweeping across the first dot matrix row of all the characters in the first display line displaying the appropriate dots along that line scan. This is then followed by matrix row 2 and so on. The video signal thus consists of the first row of matrix dots for each successive character, followed by the next row and the next up to the seventh. If for example, V F

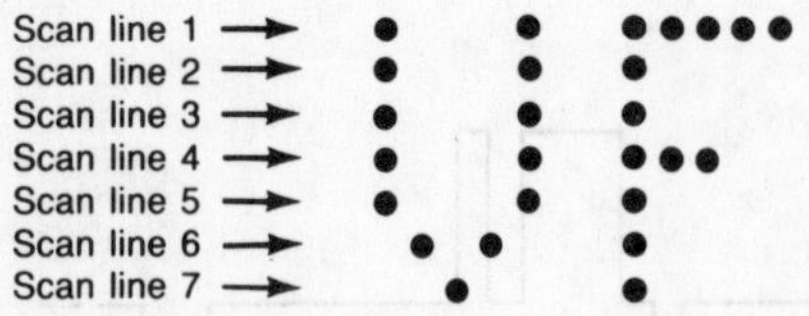

Scan line 1 →
Scan line 2 →
Scan line 3 →
Scan line 4 →
Scan line 5 →
Scan line 6 →
Scan line 7 →

Fig. 11.8

(Fig. 11.8) are the first two characters then the video signals representing the two characters (line scan by line scan) are as shown in Fig. 11.9. This video signal is generated using the arrangement shown in Fig. 11.10 where a ROM look-up chip or a character generator is used to produce a five-line output which corresponds to one row of matrix dots. The characters are selected one by one by feeding the appropriate seven-line ASCII code to the character generator. The particular matrix row to be obtained at the output of the character generator is selected by the three-line row (or line scan)

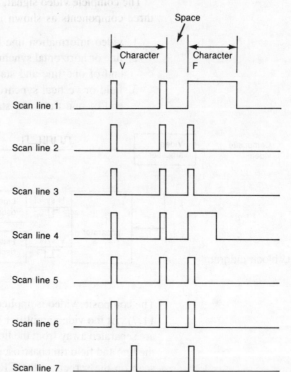

Fig. 11.9 Line-scan-by-line-scan representation of characters in Fig. 11.8

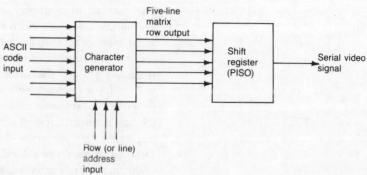

Fig. 11.10 Video generating system

address input. Thus to produce an output representing matrix row number 3, 010 is placed on the row address input. Assuming that characters V F in Fig. 11.8 are to be displayed, then the ASCII code for the first character, V, is applied to the character generator while simultaneously a 000 is placed on the row address to select the first matrix row. The character generator will then produce the dot pattern of the first matrix row of V, namely 10001. This parallel output is then converted into a serial signal using a simple shift register (or PISO) to form the first part of the line scan video signal. The ASCII code input then switches to the next character F. With the row address remaining at 000, the dot pattern of the first matrix row of the second character (11111) is produced and applied to the shift register to form the second part of the line video signal and so on for all other characters along the first display line. At the start of the second scan line, the ASCII code input reverts back to the first character (V), but this time the row address select input is incremented to 001 to select the dot pattern of the second matrix row and so on. The complete sequence of events are as shown in Table 11.1.

This process is repeated for the next display line and so on to the end of the field. The whole display is then refreshed by repeating the whole process starting with the first matrix row of the first character of the first display line and so on.

Table 11.1

ASCII code seven-bit parallel	Line select address three-bit parallel	Character gen output five-bit parallel		Shift register output waveform five-bit serial
1100110 (V)	000	10001		
1000110 (F)	000	11111	line 1	
1100110 (V)	001	10001		
1000110 (F)	001	10000	line 2	
1100110 (V)	010	10001		
1000110 (F)	010	10000	line 3	
1100110 (V)	011	10001		
1000110 (F)	011	11100	line 4	
1100110 (V)	100	10001		
1000110 (F)	100	10000	line 5	
1100110 (V)	101	01010		
1000110 (F)	101	10000	line 6	
1100110 (V)	110	00100		
1000110 (F)	110	10000	line 7	

Interlacing

With interlacing, each complete picture is composed of two fields with each field representing half a picture. When characters are displayed, each dot matrix row is repeated, the first time in the odd field and the second time in the even field. Each dot matrix row thus occupies two line scans, giving a total of 14 lines per character (or 20 lines if spaces between display lines are included).

Memory mapped screen

Refreshing the display implies the necessity of a refresh memory in which the contents of the display are stored. Most microcomputers make use of memory mapping technique with some of the system's memory (typically 8 k) set aside to hold the characters normally in ASCII format. A memory map is the addressing plan of the memory area assigned for storing the contents of the display. Each character position is given an individual memory location as shown in Fig. 11.11. Each location contains the ASCII of that character which is addressed by a controller as the electron beam scans the screen. The video memory may also be updated as the display is changed.

The CRT controller (CRTC)

Control and timing of the video generating system is provided by the cathode ray tube controller, CRTC. The CRT controller has the following functions.

1. It provides the sequence of character addresses that need to be refreshed to the refresh memory.
2. It provides the sequence of row select addresses to the character generator.
3. It generates the line and field sync. pulses for the CRT.
4. It generates all the necessary timing and clock pulses of the system.

A crystal oscillator is used as a master oscillator. Its frequency is known as the **clock** which is the rate at which successive dots are displayed. The actual dot clock frequency depends on the display layout. For a 40-columns-by-20-rows display a dot frequency of 6 MHz is typical. This frequency allows for blank margins at the top, bottom and the sides of the display. The other clock pulses including the sync. and blanking pulses are then derived by subdividing the master dot clock.

In addition to the above, two other functions are normally allocated to the CRT controller.

1. **Cursor output**. A cursor is an independent character pointer such as a blinking square or an arrow whose position is controlled by special keys or by command.

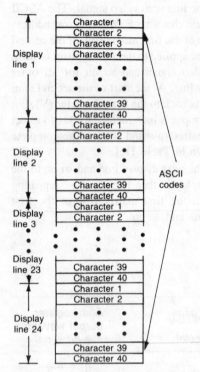

Fig. 11.11 Memory mapping of a 40 × 24 VDU display

2. **Light-pen output**. A light pen senses the light emitted by the beam as it passes. By using the time relationship to the start of the field, the CRT controller can calculate the position of the light pen on the screen.

Examples of CRTC chips are Motorola's 6845, Intel's 8275 and National Semiconductor's 8350.

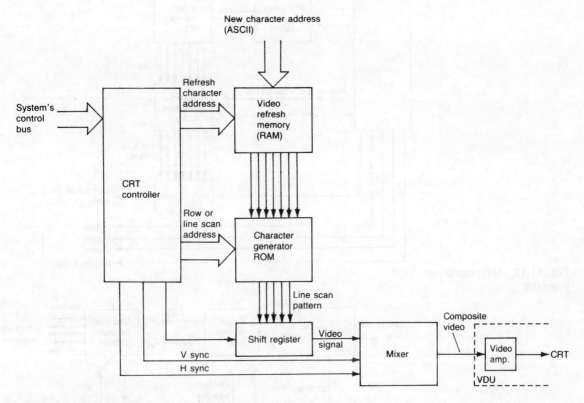

Fig. 11.12 VDU interface

VDU interface

The block diagram in Fig. 11.12 shows the relationship between the various units of a VDU interface. The two inputs to the refresh memory (the refresh and new character addresses) are normally multiplexed (as shown in Fig. 11.13) to share out the access to the memory. The CRT controller keeps tracks of the position of the beam through its clock pulses instructing the character generator to produce the correct matrix dot pattern for each line scan. The shift register is clocked by the dot clock frequency to control the dot rate of its output which forms the video signal. This video signal is then mixed with the two sync. pulses to obtain the composite video signal for the VDU.

A microcomputer–VDU interface is shown in Fig. 11.13. The

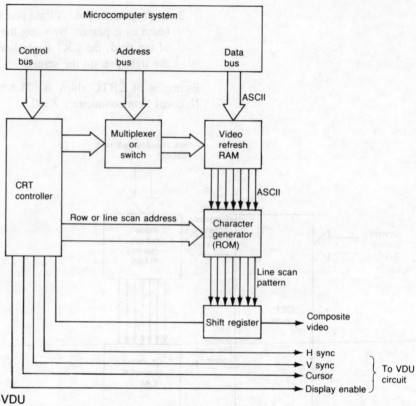

Fig. 11.13 Microcomputer–VDU interface

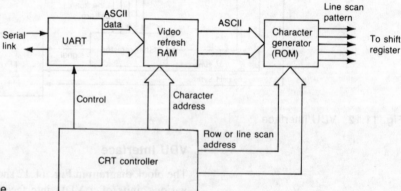

Fig. 11.14 VDU serial interface

multiplexer allows the CRT controller access to the video memory in the first half of the clock cycle and the microcomputer processor in the second half. The multiplexer could be a simple address switch alternating between the address bus of the microcomputer and the refresh character address from the controller.

Where a serial link is used to feed the data to the VDU, a UART is used to convert the serial into parallel data as shown in Fig. 11.14. In this case the CRT controller will control the status etc. of the UART to ensure that the parallel ASCII data are entered into the correct address location of the video refresh memory.

Use of domestic TV receivers

As mentioned earlier, domestic TV receivers are often used for video display purposes. The ordinary television set receives a modulated UHF signal via an antenna. The first section of the receiver therefore consists of the circuitry necessary to retrieve the composite video from the modulated UHF signal as shown in Fig. 11.15. The tuner receives

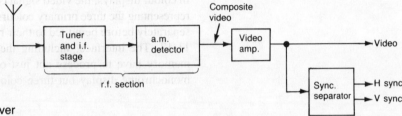

Fig. 11.15   Domestic TV receiver

the modulated UHF signal via the antenna. Following several stages of amplification and frequency conversion, the modulated signal is then applied to an amplitude demodulator or detector. The detector retrieves the composite video and feeds it to the video amplifier to be processed as described earlier. There are two ways of using a TV receiver for VDU purposes. The first is to bypass the r.f. section of the receiver and feed the composite video from the interface directly to the video amplifier. The second more popular method is shown in Fig. 11.16. In this case the composite video from the interface is first modulated on a UHF signal and then applied to the aerial socket of the TV receiver. It is then demodulated by the receiver before being applied to the video amplifier. The first method utilises a larger bandwidth, thus resulting in better definition of the display.

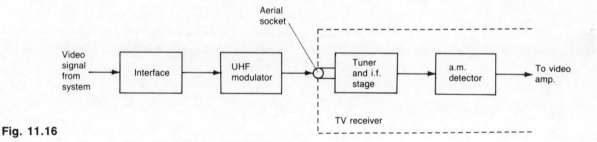

Fig. 11.16

Colour display

Colour monitors use three separate electron beams which are focused to impact upon a specially coated screen to give three separate colours: red, green and blue (R, G, B). The three colours are indistinguishable by the naked eye which sees three colours mixed together. These three colours are known as **primary colours** from which a variety of different colours may be produced (such as yellow by mixing red and green) as illustrated in Table 11.2. White contains all colours and may then be produced by adding together all three colours.

$$R + G = \text{yellow}$$
$$R + B = \text{magenta}$$
$$G + B = \text{cyan}$$
$$R + G + B = \text{white}$$

In colour displays, the video signal consists of three component parts representing the three primary colours. The three signals are amplified separately before being fed to their respective guns as shown in Fig. 11.17. The interface, including the CRT controller and the video memory have to process not just one signal as was the case of a monochrome display but three colour signals.

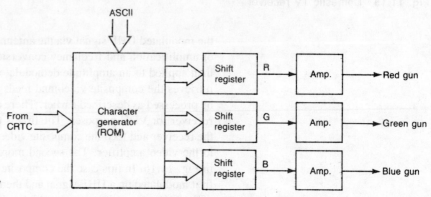

Fig. 11.17 Colour display

12 Test instruments and fault finding

There are three types of test instruments: the analogue type with analogue read-out (pointer-and-scale) such as the AVO multimeter, the analogue type with a digital read-out, normally referred to as digital meters (e.g. the DVM) and the digital state testing instruments. The first two carry out measurements on analogue signals while the latter type tests for the logic state of a test point or a node. There are a number of logic state testers such as the logic probe, the logic pulser and the current tracer. For testing bus-structured systems such as microprocessor-based systems, more sophisticated instruments are necessary, such as the logic and signature analysers.

The logic probe

The logic probe is a logic-testing instrument which investigates the logic state of a node in a digital circuit. It can indicate the presence of a logic 1, logic 0 or an open circuit (o/c) node. Two indicator lamps are used to indicate a high or a low. Open circuits or indeterminate states (between threshold levels) are indicated by no light while pulses are indicated by a flashing light. A **pulse-stretching** technique is employed by which short-duration pulses may be detected by 'stretching' them to ensure that they last long enough for the indicator to be observed by the human eye. Pulses as narrow as 10 ns may be stretched to as much as 50 or 100 ms.

The logic pulser

The logic pulser is used to stimulate gates, flip-flops, counters or other logic ICs. It drives an IC pin or node into its opposite logic state, i.e. it drives a low node high, and a high node low. Together with a logic probe it may be used to verify the function of a gate, a counter or other digital devices. It may also be used to test the continuity of bus lines. Short-duration pulses are employed in order to avoid what is known as **back-driving stress** associated with pulsers.

147

Application to digital gates

In digital applications the most common faults are what are known as stuck-at faults, namely stuck-at-one (IC pin or node permanently at logic 1) and stuck-at-zero (IC pin or node permanently at logic 0). Other faults may also occur including open-circuit pins or short circuits between tracks. Testing a logic gate takes the form of stimulating the inputs using a logic pulser and observing the effect on the output using a logic probe. For example, consider gate G1 in Fig. 12.1. First the logic states of the two inputs and the output are noted using a logic probe. The results in Table 12.1 were obtained. Although the logic levels correspond to the truth table (Table 12.2), the gate could still be faulty and further tests must be carried out.

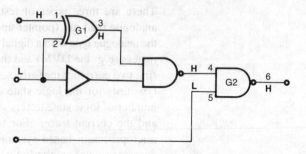

Fig. 12.1

The pulser is placed at pin 1 and the probe at output pin 3. The pulser is operated while the probe is observed for a flicker to indicate a change in the output level. It can be seen from the truth table for EX-OR gate G1 (Table 12.2), that a change in the logic level of pin 1 (making both inputs logic 0) should produce a change in the output indicated by a flicker on the probe. If a flicker is not observed, then either pin 1 or pin 3 is stuck at one. In either case a second test must be carried out. The pulser is now moved to the other input, pin 2. With the probe remaining at output pin 3, the pulser is operated. The truth table indicates that a change in the logic level of pin 2 (making both inputs logic 1) should change the output level. A flicker on the probe, therefore, indicates no fault at pins 2 or 3. The absence of a flicker on the probe indicates that either pin 2 or pin 3 is stuck at zero. If the first test produced a flicker indicating a sound pin 1 and pin 3, then the absence of a flicker in the second test indicates a stuck-at-zero fault at pin 2. On the other hand, no flicker in the first test and a flicker in the second indicates a stuck-at-one fault at pin 1. The absence of a flicker in both tests indicates that output pin 3 is stuck at one.

In the majority of cases, only one test may be applied usefully to a single gate. Depending on the kind of gate and the logic levels of its various terminals, a second test on the same gate may prove unproductive. For instance, consider gate G2 in Fig. 12.1. The logic levels of its input pins 4 and 5 and output pin 6 are 1, 0 and 1 respectively. It being a NAND gate, applying a pulser to input pin 4 to change its logic level from high to low (making both inputs low), will

Table 12.1

	Inputs		Output
Pin no.	1	2	3
Logic level	1	0	1

Table 12.2

| Inputs | | EX-OR | NAND |
A	B	function	function
0	0	0	1
0	1	1	1
1	0	1	1
1	1	0	0

not produce a change in the output whether or not the gate is faulty. This is because, as can be seen from the truth table (Table 12.2), a logic 0 to both inputs will keep the output unchanged at logic 1. The only test that may be carried out on this gate is to apply the pulser to pin 5 and the probe to pin 6. Changing pin 5 from low to high should result in a flicker at the probe indicating a changed output level provided both pins 5 and 6 are sound. A further test may now be carried out on a following or preceding gate for further diagnosis.

Once a pin is suspected to have a stuck-at fault, the fault may be confirmed by placing both pulser and probe on the suspected pin or node. If the operation of the pulser does not result in a flicker on the probe, then the fault is confirmed.

EXAMPLE

Figure 12.2 shows a part of a digital circuit. Using a logic probe. The following logic states were observed.

> Gate G1: pin 1 high; pin 2 low; pin 3 high.
> Gate G2: pin 6 high; pin 7 high; pin 8 low.

Using a logic probe and pulser the tests in Table 12.3 were carried out and the results given therein were obtained.

Identify all possible faulty pins and the nature of the fault.

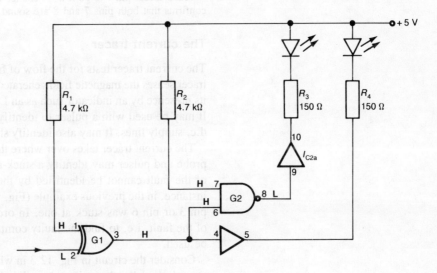

Fig. 12.2

Table 12.3

	Action	Results
Test 1	Probe to pin 3 Pulser to pin 2	No change observed on pin 3
Test 2	Probe to pin 8 Pulser to pin 6	No change observed on pin 8
Test 3	Probe to pin 8 Pulser to pin 7	Change observed on pin 8

Table 12.4

Inputs		G1	G2
0	0	0	1
0	1	1	1
1	0	1	1
1	1	0	0

Solution

The first step is to write down the truth tables for gates G1 and G2 being tested (Table 12.4). Then write the observed logic levels on the circuit diagram as shown.

Test 1 forces pin 2 of G1 to change from low to high. Output pin 3 should suffer a change, resulting in a flicker on the probe. The absence of a flicker indicates a possible fault on pin 2 or pin 3.

Test 2 forces pin 6 of G2 to change from high to low. This should result in a change in output pin 8. No change indicates a possible fault on pin 6 or pin 8. Assuming there is one single fault, then pin 3 of G1 or pin 6 of G2 must be stuck at one.

Test 3 forces pin 7 to go low. A change is observed on the output which confirms that both pins 7 and 8 are sound.

The current tracer

The current tracer tests for the flow of fast-rising current pulses. The tracer senses the magnetic field generated by these pulses and indicates its presence by an indicator such as an LED or a constant-tone sound. It may be used with a pulser to identify short circuits to earth or to d.c. supply lines. It may also identify shorts between nodes or lines.

The current tracer takes over where the probe leaves off. The logic probe and pulser may identify a stuck-at node. However, the cause of the fault cannot be identified by the probe and pulser test. For instance, in the previous example (Fig. 12.2) it was found that either pin 3 or pin 6 was stuck at one. In order to find the precise cause of the fault, i.e. to find the faulty component, a current tracer must be used.

Consider the circuit in Fig. 12.3 in which the input to U_5 is shorted to earth. Initially, the logic probe will indicate a stuck-at-zero condition at the output of U_1 and all along the track feeding the four gates U_2 to U_5. To identify the actual faulty pin, a pulser is used in conjunction with a current tracer. The following procedure is carried out with the d.c. supply **switched off**. The pulser is placed at any point along the shorted track, say node A. By pulsing this node, a current path is created as shown which ends at the faulty node F. The tracer is then used to follow the current path to its end. The tracer is first placed at node A. The presence of a current is indicated by the tracer by a lighted lamp (or an audible sound). The current tracer

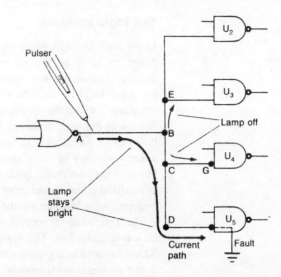

Fig. 12.3

is then moved from node A to node B. The tracer indicating lamp remains on. If the tracer is now taken towards node E, the indicating lamp will be off since there are no current pulses between B and E. The tracer is moved towards C, maintaining the indication at a constant level. If the tracer is now moved towards G, the indication will once again cease, and so on until the tracer reaches node F with nowhere else to go. In general the tracer is moved around the circuit along a path that keeps the indicating lamp (or sound) constant until the faulty node is found.

Similar procedures may be followed for a solder bridge fault and such other faults as V_{CC} to ground faults.

The logic clip

The logic clip is another device that can indicate the logic state of an IC pin. In this case the IC clip reads simultaneously the logic states of all pins. The state of each pin is indicated by an LED: on for logic 1 and off for logic 0. A pulse is indicated by a dimmed light.

The logic clip may be used to verify the truth table of an IC and to test a counter or a shift register for faults on outputs, resets, clears or other pins.

The logic comparator

The logic comparator compares the logic state of an IC pin with the logic state produced by a known-good IC. It repeats this test for all the pins and displays any errors in performance pin-by-pin. It carries out this test while the suspected IC remains in circuit (**in-circuit test**) very quickly and efficiently. The disadvantage of this type of logic state tester is the fact that a known-good IC must be available.

The logic analyser

Logic state testing instruments, described above, such as the logic probe or IC clip, are of extremely limited use when testing a microprocessor bus-structured system. In a bus-structured system, the logic level of an individual bus line does not provide enough information about the operation of the system. For this, it is necessary to look at the logic levels of all lines on the bus simultaneously. In this way, a word, such as an address on the address bus or data on a data bus, may be ascertained. Furthermore, one single word on a bus does not provide adequate test information about the system. What is required is the sequence of words that appears on the bus, e.g. a sequence of addresses on the address bus.

The logic analyser records a series of logic levels at several points on a unit under test. The logic analyser thus has several inputs (24, 32, 48 or more being quite usual) known as **channels**. The input data is not recorded continuously, but is sampled by a clock and stored in memory.

Figure 12.4 shows a basic block diagram for the triggering section of a logic analyser. The input channels are connected to the appropriate test points, e.g. in an eight-bit microprocessor system, 16 channels may be connected to the address bus, eight channels to the data bus and others to the control lines. The input data is fed into the data latch, where the logic levels of each channel are held long enough for the analyser to capture the information accurately. This data is fed into the memory block which stores the data for later display. The memory block is activated to capture and store the data by the write control from the trigger word detector. The trigger word detector is preloaded with a **trigger word** representing a combination of the simultaneous logic states of the input channels. The stream of input data is fed into the trigger word detector as shown. When the detector recognises the preloaded trigger word occurring in the input channels, it enables the

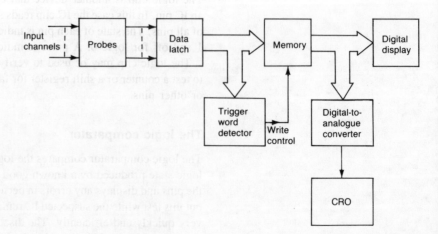

Fig. 12.4 Logic analyser

write control. The memory then stores a block of data activity before and after the occurrence of the trigger word. The captured data may then be displayed in binary or hex format.

Timing analysis

Once the input data has been captured and stored in memory, it may be displayed as logic levels in the form of waveforms for **time-domain** analysis. Timing analysis involves displaying simultaneously the waveforms of the input channels. These waveforms are produced from the digital data stored in the memory of the analyser by the use of a digital-to-analogue converter, as shown in Fig. 12.4. The DAC modifies the digital data to produce a multi-channel display on an oscilloscope.

Timing and sampling

As explained in Chapter 9, a microprocessor system operates on a fetch and execute cycle. The timing cycle is produced by an internal clock which synchronises the operation of the system. The logic analyser must therefore capture the information at least once every cycle of the unit under test. The input data is, in fact, sampled at a rate determined by a sampling generator, which resets the latch to receive the next batch of input data.

There are two modes of operation, namely the **synchronous** mode and the **asynchronous** mode. In the *synchronous mode*, the acquisition of data occurs synchronously with the timing clock of the unit under test.

In the *asynchronous mode*, the analyser uses an internal clock to trigger the sampling generator. The internal clock must be faster than the speed of the input data, otherwise complete logic states will go unrecorded. In this way, the logic states of the input channels between clock cycles of the system under test can be investigated. Unwanted transitions such as **glitches** may be captured provided a sampling active clock edge occurs during the glitch.

The qualifiers

In fault finding on a microprocessor-based system, it is essential to be able to discard irrelevant data and capture only the data surrounding the fault. By the use of a trigger word, the captured data is reduced to those occurring before and after the preloaded trigger word. However, in most applications, this is not sufficient, and further selection is necessary. Also, in cases where the trigger word occurs more than once in a sequence of events, the analyser must be able to distinguish and choose the required trigger event. To overcome both these problems, trigger qualifiers and clock qualifiers are used.

Trace control

Another way of controlling the recording of data is by means of 'trace' words which function in a similar way to the clock qualifier. A trace word defines the condition for the data input which must be satisfied, e.g. channel 1 to be high and channel 5 to be low for the analyser to capture and store the input data.

Signature analysers

Logic analysis is employed mainly in design and development, although programmable logic analysers may be used in production testing as part of an automatic test system.

In fault-finding, logic analysers suffer from the need to interpret the captured data, which means a prior knowledge of the workings of the circuit. Skilled manpower is, therefore, necessary. The signature analyser is designed to overcome this problem by allowing unskilled labour to diagnose and repair most faults in a system.

Principle of signature analysis

The introduction of digital electronics rendered the concept of signal tracing impractical as a method for fault diagnosis. In analogue electronics, an immediately recognisable waveform with specified frequency and amplitude is present at each test point as shown in Fig. 12.5. By comparing actual waveforms present on a system with those expected on a good system, the test engineer can identify a fault. In digital electronics, this procedure is not possible since trains of pulses representing binary codes look very similar to each other. The equivalent of an analogue waveform at a test point is a compression of a stream of bit pulses at a node into a recognisable word such as a hexademical number, known as a **signature**.

The simplest type of signature analyser is that shown in Fig. 12.6. The start signal begins the process of capturing the data bits in the shift register. This process is terminated by a stop signal into the control. The data bits captured by the shift register are then displayed as a four-digit hex or other coded signature. As may be seen in Fig. 12.7 with the control signals 'start' and 'stop' as indicated, the data captured by the signature analyser are as shown. The shift register transfers the bits from left to right at the selected edge of the clock waveform (negative edges in this case). Thus, following the start

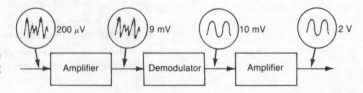

Fig. 12.5 Analogue schematic with expected waveforms at various test points

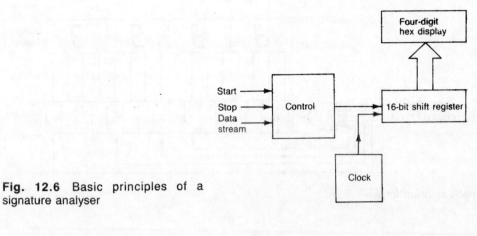

Fig. 12.6 Basic principles of a signature analyser

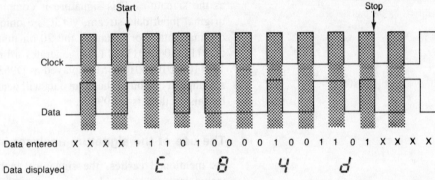

Fig. 12.7 Timing diagram for a signature analyser

trigger edge, the shift register begins to sample and capture the data bits at every negative clock edge. This continues until the gate is closed by a 'stop' trigger edge. The data entered and displayed is therefore 1110100001001101 = E84D(hex). This simple signature analyser suffers from an obvious disadvantage, namely it cannot handle a data stream longer than 16 bits. After the 16th bit, data bits will begin to be lost with only the last 16 remaining in the register to be displayed. This may be overcome by expanding the shift register to 24, 32, 64, ..., bits which produces signatures of 6, 8, 16 ... hexadecimal digits. Such a method is impractical and, even for a 64-bit data input, is expensive. A clever way of overcoming this limitation is the use of **cyclic redundancy count**, employing four feedback loops in the shift register. This ensures that data bits longer than 16 are recirculated producing a unique signature of four digits only (see Fig. 12.8). The feedback loops are connected to the input of the register via the exclusive-OR gate shown. In this way the bits are not lost but merely circulated round the register continuously updating its contents. At the end of the measurement when the stop trigger edge closes the gate, a residue remains in the register which is displayed

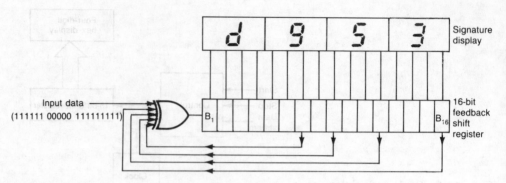

Fig. 12.8 Basic block diagram for a signature analyser

as the signature. This signature is completely dissimilar from the original input data stream, yet it is a unique representation of that original data. For instance, the 20-bit input sequence in Fig. 12.8 of 11111100000111111111 generates a unique 16-bit signature of 1101100101010011(bin) displayed as D953(hex). The certainty that a change in one bit of the input data will generate a different signature by this method is 99.998%.

The use of the signature analyser in fault finding

As mentioned earlier, the signature analyser opens the way for technicians to test and repair a digital system without detailed knowledge of the circuit. In order to do this, test programs must be written to test the various parts of the system such as the CPU and memory chips. The system is then stimulated with the test program and a signature analyser is used to test for correct signatures at predetermined test nodes. The test program may be stored in ROM in the system under test itself, or it may be fed in externally by a peripheral device or by the signature analyser itself. The signature at each node is monitored and compared with that produced by a known-good system. The analyser may have a facility to store the known-good signatures to make immediate comparisons. Any incorrect signature indicates a fault. This fault may be identified by '**signature tracing**' which involves testing other nodes to identify the faulty block.

A test program for checking ROM devices may involve the data contents of each location. That for RAM may involve writing a 1 or a 0 in each location and then proceeding to read the contents of each location. This is known as **checker board test pattern**. For testing either a ROM or a RAM chip the contents of all locations are fed into a signature analyser to produce a known-good signature. Any other signature that is obtained indicates a faulty chip. Testing the CPU involves going through some or all of its instructions and checking the response.

The test program includes the start and stop signals for the signature analyser. In the absence of a dedicated test program the start-up program of the microprocessor system itself is normally used. The microprocessor RESET signal may be used to start the analyser. A stop signal can be obtained by dividing the system's clock by a suitable factor, e.g. 100 as shown in Fig. 12.9. Assuming a clock frequency of 1 MHz, then the analyser will capture a stream of pulses appearing at a node for a period of 100 μs. This method of selecting the start and stop signals is most suitable for detecting a stuck-at fault in a microprocessor system. It should be noted that while the signature for a stuck-at-0 node is 0000, that for a stuck-at-1 is determined by the time interval between the start and stop signals.

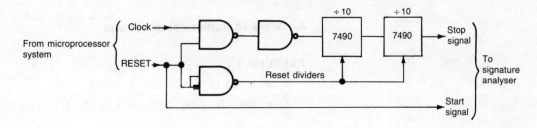

Fig. 12.9

Appendix

Answers to numerical questions

EXERCISE 1.1

1. (a) 6   (b) 14   (c) 21   (d) 45   (e) 63.
2. (a) 101_2 (b) 10001_2 (c) 101010_2 (d) 11111_2 (e) 101111_2.

EXERCISE 1.2

1. (a) 26_{10} (b) 47_{10} (c) 139_{10} (d) 110_{10}.
2. (a) 34_8 (b) 210_8 (c) 537_8 (d) 1253_8.
3. (a) 010 111   (b) 010 001 000   (c) 101 101 101
   (d) 110 101 100 011.
4. (a) 2_8 (b) 63_8 (c) 131_8 (d) 1270_8.

EXERCISE 1.3

1. (a) 0010 1010   (b) 1000 1101   (c) 1100 0000 1001
   (d) 1110 1111 0010   (e) 1111 1111 1111 1111.
2. (a) D6   (b) 32   (c) 97F   (d) EB53.
3. 1249_{10}.

EXERCISE 1.4

1. (a) 0.01_2 (b) 0.00111_2 (c) 0.01111_2.
2. (a) 0.25_{10} (b) 0.90625_{10} (c) 7.4375_{10} (d) 41.25_2.

EXERCISE 1.5

(a) 110001_2 (b) 100110_2 (c) 1001_2 (d) 01011_2.

EXERCISE 1.6

1. (a) 00010111   (b) 11101001   (c) 11010010   (d) 100000011.

EXERCISE 2.1

(a) Truth table

A	B	P	Q	R	S
0	0	1	1	1	0
0	1	1	1	0	1
1	0	1	0	1	1
1	1	0	1	1	0

(i) Boolean expression; EXCLUSIVE-OR   (ii) function; $A \oplus B$.
(b) One combination only; $A = 1$, $B = 1$.

EXERCISE 2.2

(a) Boolean expressions at
P: $\overline{A}$, Q: $\overline{B}$, R: $\overline{A+B}$, S: $\overline{\overline{B}+A}$, T: $\overline{\overline{A}+B} + \overline{A+\overline{B}}$.
(b) Truth table

A	B	P	Q	R	S	T
0	0	1	1	0	0	1
0	1	1	0	0	1	0
1	0	0	1	1	0	0
1	1	0	0	0	0	1

The circuit may be substituted by an EX−OR followed by a NOT gate.

EXERCISE 2.4

(a) $A + B$ (b) A   (c) $B + A.C- + A-.C$.

EXERCISE 6.2

(b) $1 k \times 4 = 4096$ bits.

EXERCISE 6.3

(b) $1 k \times 8 = 8192$ bits.

Index

MATTHEW BOULTON
COLLEGE LIBRARY